Connecting Maine's Capitals by Stagecoach

Leland J. Hanchett, Jr.

Pine Rim Publishing Falmouth, Maine

Front Cover

Thomas Hanchett, the author's third grandson in chronological order, took on the project of creating a departing stage coach scene in front of the Elm Tavern stage house in Portland, Maine. Working from a couple of old drawings and one photograph, Thomas, with his trusty computer, created a drawing that would have been the envy of the many contemporary artists who tried to portray this scene. We expect to see Thomas' work in many other venues in the near future.

Back Cover

This is the story of a stage coach born in Concord, New Hampshire in 1849, and used in Bath, Maine for ninety years. Its next stop was Litchfield Park, Arizona where it served the guest of the Wigwam Resort for another twenty years. It then drew intermittent duty at Arizona State University in Tempe, Arizona, waiting to join other historical artifacts at a proposed museum there. When the museum failed to materialize, the stage coach was purchased by the grandson of a previous owner in Bath, Maine. There it sat in near ruin for a couple of years until a museum in Maine took pity on it and had it restored to its original brilliance and more. Now, it is owned by a proud collector of horse drawn commercial vehicles in Conway, New Hampshire. We will see it at fairs and displays for many years to come, a brilliant reminder of the glory days of the old stagecoach.

Typeset in Cheltenham by Ya Ya Ya Creative – www.yayayacreative.com

Printed and bound by Walch Printing, Portland, Maine

First edition—First printing

ISBN No. 978-0-692-94135-5

Library of Congress Control Number 2017953106

PRINTED IN THE UNITED STATES OF AMERICA

This book is dedicated to Ms. Elizabeth Ring,
devoted history teacher and inspiration to her students.

Table of Contents

Preface

Around 1950, Elizabeth Ring, history teacher at Deering High School in Portland, Maine, created an aid to the teaching of Maine history for the public schools of this state. In her conclusion she clearly describes the problem that so many of today's younger citizens encounter concerning the opportunities for growth and success here in Maine. Ms. Ring states:

> There are many things in Maine that need correction. Purely aside from her policies as a state, her economy is in a continual state of flux. In both politics and business, the state needs men and women of foresight and great civic consciousness to carry on the fight against defeatism and reaction - men and women who will stick to their guns and stay in the state even though discouragements come and opportunities seem greater elsewhere.
>
> The responsibility for finding these people and training them for future service falls squarely in the lap of those of us who teach.

Having been one of Ms. Ring's students, the author of this work left Maine shortly after high school graduation to attend college in California with the intent of making his fortune "out west," and maybe someday returning to Maine. After sixty years of living with that goal in mind, the return is finally at hand.

The need for informed citizenship has never been greater. Mainers need to be proud of their state. Some of that pride will come only from the knowledge of those who have gone before us, those who stayed through good times and bad to make Maine an even better place to live and grow. The years from the Revolution to the Civil War were a formative period in Maine's history. Hopefully, this small work will provide some information on a few of the people and ideas that have helped to make the State of Maine what it is today.

Part One

Introduction to Part One

From Boston to Portland

According to the Maine State Library, the name Maine first appeared in 1622 in a charter from the Council of New England granting land to Sir Ferdinando Gorges and Captain John Mason. Captain Mason exclusively held land which he named New Hampshire. Gorges' land became the Province of Mayne by order of King Charles I.

Originally, it was thought that Maine acquired its name from the Province of Maine in France, but it turned out that Charles' I queen, Henrietta Maria, did not own the Province of Maine in France as was previously supposed. Today's prevailing theory is that it was called Maine to distinguish it from all the islands nearby. The nautical term, "Main Land," is a more probable origin.[1]

King Charles I of England had granted a charter to Sir Ferdinando Gorges for all the lands from what is now Portsmouth, New Hampshire to the city of Bath in Maine, or from the Piscataqua to the Sagadahock rivers. The few original settlements, from 1626, were soon taken over by grants from the Plymouth, Massachusetts Colony to its members.

According to Dr. Willis, in his booklet *Submission of Maine to Massachusetts*, Maine was intimidated sufficiently by the government of Massachusetts to actually submit themselves to being governed from Boston.

> On the 19th of May, 1643, the Puritan portion of the Colonies including Massachusetts, Plymouth, Connecticut and New Haven, formed a compact for mutual aid and council, believing the opportunity a good one because at this time, the House of Commons was in sympathy with them. The Province of Maine, being under Episcopal rule, could not be admitted, and New Hampshire had made an alliance with Massachusetts the year before.
>
> In 1648, the Province of Maine extended from the Piscataqua river to the Kennebec. Eastward from the Kennebec, it was known as the Province of Lygonia. In a way, these two provinces were rivals; the former under Gorges, the latter under Rigby. They were about equal in population, although Lygonia was larger in territory.

[1] Maine State Library, *Origin of Maine's Name*. Retrieved July 4, 2017, from Main.gov website: <http://www.maine.gov/msl/maine/meorigin.htm>

On July 8, 1649, after the death of Gorges; Wells, Gorgeana, Kittery and the Isles of Shoals, held a convention and after long discussion, they formed a social compact.

They elected a Governor, Edward Godfrey, five counselors, and Edward Rishworth was made recorder. This contract was continued two years.

When they heard of the death of King Charles, they indicated their willingness to take directions from Parliament, and asked by petition on December 1, 1651, for the same rights and privileges as were bestowed on the other Colonies.

Previous to this, in October of 1651, at the session of the General Court, Massachusetts planned to extend her jurisdiction over Maine because she had long felt the danger of the form of government existing there as it was so entirely different from her own. Her people had left their homes, many of them of comfort and luxury, for the enjoyment of religious freedom, while Maine's settlement was made up of those who had come to seek their fortune with no special regard for any religious principles, although they did tend to prefer the faith of the Episcopal or High Church of England

Winthrop and his government hated to see this directly opposite form of church and state so near, and they shrewdly went about to change all this and prevent the danger to themselves. They guessed that their charter, if taken literally and a survey made, would give them quite a little of Maine Territory. They voted unanimously on March 31, 1652 to send surveyors to look after this. Marks on their survey may be seen in the famous Endicott Rock, at the outlet of Winnipesaukee at the weirs on the Meredith side. There, a large boulder bears the inscription of the initials of the Commissioners and John Endicott, Governor.

The report of this survey was that Massachusetts owned as far east as the Presumpscot River, and she immediately set about to take possession. On the 14th of October, 1651, Deputy Governor Bellingham and Mr. Symonds were selected by the Massachusetts Court to draw up an address to the Governor of Maine.

The Commission was made up of her strongest men. Bradstreet was a Councilor of long experience, Dennis was the commander of the militia, and Hawthorne the Speaker of the House; but they returned unsuccessful after proclaiming to the people of Maine the right of Massachusetts to govern them. They told them that they need pay no further attention to Governor Godfrey or his officials.

In October, 1652, another commission consisting of Bradstreet, Simmons, Wiggins and Pendleton, was appointed to visit Kittery with instructions to give notice as follows: 'In the name of the Government of Massachusetts to require you, and every one of you, to assemble together before us, at the house of William Everett, between seven and eight o'clock in the morning, the 16th of this present

November to declare our just right and interest to, and jurisdiction all over the tract of land where you inhabit.'

The Everett Inn had a most memorable meeting. There were wars and rumors of wars. Not low-minded retaliations, but honest and outspoken convictions. The early men of Kittery were not shallow. With a solid wisdom and a clear comprehension of the value of the Maine seacoast, they fought against the compulsory surrender to the graspings of Massachusetts.

But in 1652, Maine had but a fringe of settlements on its shores. In Maine, Kittery was the only incorporated town.

After struggle and discussion at the Eliott (Everett) Inn, forty men and one woman, Mary Bachiller, submitted. A principle portion of Maine was under control of Massachusetts.

Nine years after Massachusetts' assumption of authority, the rule of Cromwell having come to an end and royalty restored again, Parliament pronounced the claim of Massachusetts unlawful and unfounded. Commissioners were sent, ordering directly from the King that Massachusetts give up her claim and return the Province to its rightful owners. Massachusetts refused to do this and when the King's Commissioners attempted to set up a government, sent men of her own to form a court, with orders to arrest any who disputed her authority. Constant troubles went on and conditions were most unsettled, until 1667, when Massachusetts succeeded and purchased the Charter from the heirs of Gorges, which at least gave them an undisputed title, and government became settled.[2]

A payment of 1250 pounds sterling was made by Massachusetts to the family of Ferdinand Gorges. The major controversies were settled by 1692 and the District of Maine was incorporated with Massachusetts. Thus, Maine then had its capital in Boston where it would remain for the next 128 years.

The representatives and business men from Maine who attended government sessions in Boston, as well as the general public, would need to travel many miles to reach their capital and commerce center in Boston.

Two options were available to these citizens, travel by land or by river and sea. Certainly, the latter was more acceptable considering the condition of the roads and the lack of public transportation on them. Travel by ship could be comfortable except in stormy weather. Travel by land was treacherous, coupled with the fact that there were no stages and few accommodations.

Maine faced two major hurdles between 1810 and 1820. The first was the War of 1812 with England, and the second was the separation from Massachusetts. Both took their toll, but both also set Maine on the path to independence and statehood.

[2] J. L. M. Willis, *Submission of Maine to Massachusetts.*

The campaign for separation began before the War of 1812, possibly as early as 1785. In that year, an opinion writer for the Falmouth Gazette asked the question: "Have we not good harbors and well situated for fishery, and a foreign trade, and materials for shipbuilding and the lumber business? Have we not a great tract of cultivated land with only a few roads into the wilderness? If so, how are these natural advantages to be made useful to us, or to the public, but by giving the greatest encouragement possible to people to cultivate wildland and bring the lumber to the markets? And how is it possible to accomplish these valuable ends in any other way so well as by restoring to the Province of Maine its ancient privileges?"

Conventions were held periodically in an attempt to ascertain the public feeling on the question of separation. The citizens of Boston continually reminded Mainers that they did not know how to run a government, and they certainly could not afford a separate government even if they could manage it.

By late 1815, all-out war began between two Portland newspapers. The Eastern Argus was on the side of separation, while the Portland Gazette would hear nothing of it. Each fired volleys at the opposing view with the Eastern Argus sending the first shot across the bow of the Gazette on November 18. In a series of fourteen articles, the Eastern Argus addressed the issues of sufficient population, sufficient means, sufficient talent, expense to separate, and advantages of doing so.[3]

[3] Eastern Argus (Portland, Maine) November 18, 1815 through May 7, 1816.

It took four months before the Gazette retaliated on February 13, 1816. Not surprisingly, a few gentlemen of the Republican Party were blamed for all the "separation fever." The Gazette also argued that Maine had received its current growth and prosperity because it was a part of Massachusetts. No credit was due to the Mainers themselves.[4]

The Gazette provided nine articles in their series which ended just before May 20, 1816 when the issue came down to a vote. The results were 10,393 for separation and 6,501 against. Separation was desired, but further delays would occur. Apparently, Massachusetts did not want to lose a land mass many times larger than its own.

Another explanation for the desire to separate was given as follows:

> Mainers remained a little timid until the War of 1812 broke out. Washington was plundered, New Orleans was placed under siege, but Maine suffered the most. In the fall of 1814, British forces moved down the coast of Maine, occupying some towns, looting others. Both the District of Maine and the Federal Government looked to Boston to send relief to Maine. It never happened and the popular opinion in Maine moved towards separation.[5]

It is interesting to note that the issue of lack of support by Massachusetts during the War of 1812 never came up in the newspaper articles. As so often happens, many things are involved

[4] Gazette (Portland, Maine) February 13, 1816 through April 9, 1816.

[5] Portland Press Herald (Portland Maine) June 24, 2012, Colin Woodward, *The War That Made Maine a State*. June 24, 2012.

in any decision for action. A single event is rarely sufficient to create big changes.

It took until 1819 for the Massachusetts Legislature to give the question due consideration. They finally agreed that it was the best course and a public referendum was conducted. The result was overwhelmingly in favor of separation, and in October of that year, a convention was called to establish a State Constitution. It still took a compromise in the Federal Government to secure the statehood that so many Mainers wanted. Missouri had to be admitted as a slave state for Maine to be admitted as a free state. The Congress of the United States created the State of Maine on March 16, 1820, and the first State Legislature convened in Portland on May 31, 1620. Portland was now the new capital.[6]

Meanwhile, New Hampshire was having its own problems in its relationship with Massachusetts and maintaining its provincial capital in Portsmouth.

Wikipedia tells us that:

> The relationship between Massachusetts and the independent New Hampshirites was controversial and tenuous, and complicated by land claims maintained by the heirs of John Mason. In 1679 King Charles II separated New Hampshire from Massachusetts, issuing a charter for the royal Province of New Hampshire, with John Cutt as governor. New Hampshire was absorbed into the Dominion of New England in 1686, which collapsed in 1689. After a brief period without formal government (the settlements were de facto ruled by Massachusetts) William and Mary issued a new provincial charter in 1691. From 1699 to 1741, the governors of Massachusetts were also commissioned as governors of New Hampshire.
>
> Because New Hampshire's governorship was shared with that of Massachusetts, border issues between the two colonies were not properly adjudicated for many years. These issues principally revolved around territory west of the Merrimack River, which issuers of the Massachusetts and New Hampshire charters had incorrectly believed to flow primarily from west to east. In the 1730s New Hampshire political interest led by Lieutenant Governor John Wentworth were able to raise the profile of these issues to colonial officials and the Crown in London, even while Governor and Massachusetts native Jonathan Belcher preferentially granted land to Massachusetts interests in the disputed area. In 1741, King George II ruled that the border with Massachusetts was approximately what it is today, and also separated the governorships of the two provinces. Benning Wentworth in 1741, became the first non-Massachusetts governor since Edward Cranfield succeeded John Cutt in the 1680s.[7]

6 William Willis, *The History of Portland.*

7 Wikipedia, *History of New Hampshire.* Retrieved July 4, 2017, from Wikipedia.org website: <https://en.wikipedia.org/wiki/History_of_New_Hampshire>

According to R. Stuart Wallace, Director, Division of Historical Resources for the State of New Hampshire:[8]

> By the time of the American Revolution, New Hampshire was a divided province. The economic and social life of the Seacoast revolved around sawmills, shipyards, merchant's warehouses, and established village and town centers. Wealthy merchants built substantial homes, furnished them with the finest luxuries, and invested their capital in trade and land speculation. At the other end of the social scale, there developed a permanent class of day laborers, mariners, indentured servants, and even slaves.
>
> In the central and western parts of the province, however, the inhabitants were farmers. Many, if not most, had come from Massachusetts, Connecticut, and Northern Ireland, and their ties to Portsmouth were weak. They spread themselves over the countryside, clearing small lots and building simple one- and two-story farmhouses. Their towns were punctuated with a few sawmills and gristmills, a number of taverns, a meetinghouse, and perhaps a store or public school. During the War for Independence, farm towns in the Connecticut [River] Valley became so disenchanted with provincial leaders in the seacoast area that they attempted to secede. Only by agreeing to hold several legislative sessions in the Merrimack Valley, and particularly in the town of Concord, did New Hampshire keep itself from breaking apart.
>
> New Hampshire's sectionalism was intensified by its mountainous terrain. The state's major rivers run from north to south, which dictated that most roads, and later railroads, would lead to Boston, rather than Portsmouth. New Hampshire's economic and social ties to Boston also meant that its residents were more sensitive to events in Boston. When Revolutionary politics closed the port of Boston, New Hampshire sent food. When news came of Lexington and Concord, New Hampshire towns sent troops. The once-loyal British subjects of New Hampshire became the first in America to draft a separate state constitution, and they were the first to instruct their delegates attending the Continental Congress in Philadelphia to vote for independence.
>
> Throughout the 1800s, the Seacoast declined as a commercial center. As economic activity slowed in Portsmouth, towns like Dover, Newmarket, and Somersworth prospered by turning to textile manufacturing. It was the Merrimack Valley, however, that took over as the social, political, and economic center of the state. Manchester and Nashua became major textile manufacturing centers, while Concord's central location and diversified economy made it well-suited to serve as the new state capital.

[8] New Hampshire Historical Resources. Retrieved July 4, 2017, from NH.gov website: <https://www.nh.gov/nhdhr/markers/brief.html>

So, Portsmouth was the capital of New Hampshire until it was not! That occurred just as the stage lines were beginning to take hold in 1808.

In Part One, we will examine the effort which was required to build roads and bridges, form stage lines, and provide suitable accommodations for the traveling public who needed or wanted to get from Portland or Portsmouth to Boston and return.

Chapter One

Building the Roads from Boston to Portland

The Revolutionary War was over and the colonies needed to get organized. Providing postal service was an important step and the existing Post Roads were nothing to be proud of! Northern New England was still a frontier and most transportation was by river and sea. Maine was just a district of the Massachusetts Bay Colony, separated from its capital, Boston, by a small stretch of New Hampshire.

Enterprising New Englanders began thinking about toll roads or turnpikes as they were called then. The name came from the pike or barrier placed on private roads which when turned allowed public passage once the toll was paid. These far-sighted individuals knew that travelers and freight would need to pass efficiently from inland resources to seaports if commerce was to be encouraged.

An article from the Portsmouth, New Hampshire Gazette from 1800 laid out the task at hand:

> **Turnpike Roads**
>
> It would be very desirable to have straight and good roads throughout the United States, from one town or city to another; but particularly so from the capital to that of another. Boston being not only the capital of Massachusetts, but also of New England we ought to use our utmost exertions to facilitate the travelling to it, by the shortest possible way. It has for some time been in contemplation to open a turnpike road from Boston to Portsmouth in or very near a direct line, which would cross Merrimac river near Hoyt's wharf, and which would contract the travelling from 68 to about 52 miles, beginning the reckoning at the old State House in Boston. However great the undertaking may appear to some, by many it is considered as very trifling when compared with its immense utility to the public. In New Hampshire, a turnpike has lately been opened from the Piscatagua bridge to Concord of about 30 miles at the expense of about $900 per mile, and a petition is now before the legislature of that state for opening a road from Portsmouth to the dividing line between the states, and it is said that

> gentlemen have offered to contract with the proprietors to complete it at $1100 per mile, should it cost even $2500 per mile from hence to Boston, it would, considering its importance in point of utility and convenience, be perhaps as cheap a bargain as a public ever made.[1]

The first step was to cover that expanse from Boston to the Merrimack River in the state of Massachusetts. The New Hampshire Gazette tells us that:

> The post road from Portsmouth to Merrimack Bridge has lately been accurately surveyed at the expense of a few gentlemen in Portsmouth, by which means the exact distance is found to be 22 miles and 91 rods; and upon a straight line but 18 miles. The road from Newburyport to Malden Bridge has also been surveyed and the distance upon a straight line found to be but 27 miles. And it is well ascertained that there may be a good road from Portsmouth to Boston, and the distance but about 48, or 49 miles. Whereas the road now traveled is as much as 57, consequently there will be a full saving of nine miles; more than 1/6 the distance as now traveled.
>
> This at first may seem almost incredible, but the surveys in both instances have been made by very accurate surveyors, and the fact is well demonstrated. There requires nothing but a little exertion, to have an excellent turnpike road from Portsmouth to Boston and the distance at most under 50 miles.
>
> It is also said that the post road from Portsmouth to Portland may be considered shortened and the distance reduced to 50 miles. We do not assert this to be fact; but it seems by no means impossible. May we not therefore expect that in the course of a few years, the mail will be transported from Boston to Portland, in the same length of time it is now, from Boston to Portsmouth? It is true this will not easily be performed by those who are attentive to try how long they can make an old horse run in the stage, but the thing of itself is quite feasible. The mail stage leaves this town before 3:00 AM and arrives at Boston between 6:00 and 7:00 PM; making about 16 hours upon the road. And the whole of the time of resting will not amount to 2 hours; so that the rate of traveling is under 5 miles an hour. A mail stage traveling it [Post Road] should be called the flying mail!!
>
> At the rate of 8 miles an hour, 12 hours and 1/2, would reach Portland from Boston provided the distance should have reduced to 100 miles.
>
> Allowing 1 hour for crossing Portsmouth Ferry, there would be 2 hours and 1/2 for resting upon the road. And if the ways are good, 8 miles an hour is easily traveling, for horses that are suitable for running in the stage, if they are not drove too far without changing: especially if the mail carriers are restricted to six passengers, as they certainly ought to

[1] New Hampshire Gazette (Portsmouth, New Hampshire) March 23, 1802 p1.

be. Should a regulation of this kind be brought about (and there is no impossibility in the thing even if the Portland road cannot be so much shortened) we shall see the time when a merchant in Boston may write to his correspondent in Portsmouth one evening and received an answer to his letter of the next. We hope there are friends to the eastward [Maine] who will take the hint, and make the necessary examination and surveys. No people are more interested in the business than they are, the great increase of population, commerce and wealth, in that country, requires that the experiment should be made, and there can be no doubt but success will crown the attempt. It is most devoutly to be wished, that no mean selfish views, will mar so laudable an undertaking. 'That it will not go by Uncle Josiah's barn; or Cousin Joseph's pigsty; or may possibly spoil sister Tabatha's potato yard; or that the Col. or Capt. whose children love mightily to see the fine folks ride along in their chronicles and phaetons may object, because the road does not go by their house.' These are generally the cogent reasons to prevent laying out new roads to benefit the public. And we hope that those who have already become rich by commerce, will not withhold their money, but will vest it in a fund, not only certain and productive but whose beneficial influence will be felt by all classes of citizens, the man of pleasure not excepted. To our political fathers a word may, with propriety, be addressed.

The prejudice against turnpike roads seems to be pretty well worn off. It is now a conceded fact; that those who use the roads are the persons who ought to make it and keep it in repair; and it is expected that liberal grants will be made; for without such, the work will never be effected. How pleasing the inflection that while commerce with expanded wings brings her blessings from all quarters of the globe; the attention and industry of man, is rendering them to make communication through our country more cheap, expeditious and delightful, and so greatly increasing the value of those blessings! Consider this well ye sordid of every description, and you will neither prevent nor discourage the making of turnpike roads and bridges but if you are rich you will abundantly contribute your riches for promoting such beneficial undertakings.[2]

Boston to Newburyport

The story of the Newburyport Turnpike is one of success and failure. Success, because it was designed and constructed with precision and efficiency. Failure in that after a pittance of dividends, the owners realized a loss of 98% on their original investment. As a private road, it only lasted forty years but during that time, it moved people and goods between Maine, New Hampshire, northeastern Massachusetts, and Boston leaving a legacy that endured for 122 years. The road was intended to be the straightest route from Chelsea Bridge in Boston

[2] New Hampshire Gazette (Portsmouth, New Hampshire) November 25, 1800 p3.

Newburyport Turnpike[3]

Newburyport Turnpike Over a Hill-Topsfield Tollhouse[4]

3 Wallace Nutting, *Massachusetts Beautiful.*

4 Postcard dated 1913.

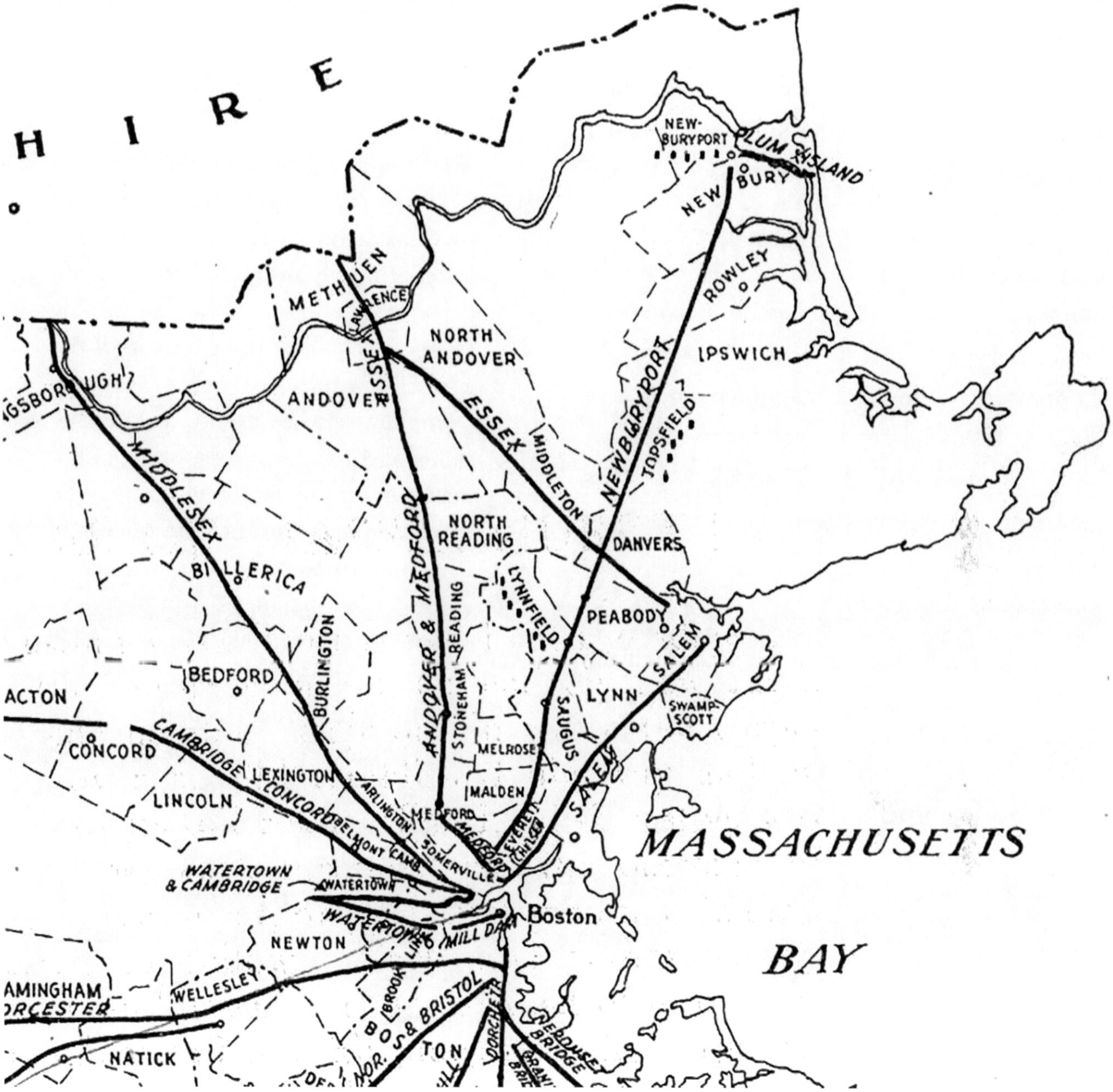

The Newburyport Turnpike[5]

to the center of Newburyport. Much like our turnpikes of today, it was the shortest route with the minimum grade possible at the time. Over hills, through valleys, and across streams it went, and most importantly, it avoided all towns except for Boston and Newburyport.

The bill for its creation was signed by Governor Strong on March 8, 1803 with the acclaimed purpose of connecting the town of Newburyport with the capital of the state, Boston. The owners of the corporation first met in April of 1803.[6] By September of that year the first bids were in for construction of a twenty-mile section south from

[5] By F.J. Wood (Scanned from *The Turnpikes of New England*) [Public domain], via Wikimedia Commons.

[6] George A. Kyle, *The Straight Road.*

Newburyport.[7] A year later the owners petitioned the Legislature for permission to build a new bridge over the Charles River in line with Barton's Point, in order to reduce the total length of the turnpike to thirty miles.[8]

Construction proceeded with haste and as sometimes happens, a man was severely injured while dynamiting ledge two miles south of Newburyport.[9] Just twenty months after the corporation's founding, an announcement was made in a Boston newspaper that twenty-five miles of road had been completed. This included six bridges and two houses of entertainment, or taverns. Completion was expected by that spring.[10]

An article in a Worcester paper, dated January 30, 1805, featured the success of the road to that point.[11]

> **Newburyport Turnpike**
> The fruit of enterprise as it relates to public roads, bids fair to be of great utility. The following particulars respecting the laudable undertaking of the proprietors of the turnpike road from Newburyport to Boston, cannot fail to be acceptable to our readers. The account is communicated by a correspondent who lately made a tour in that part of the country, and who was so well pleased with the road, that he applied to some of the contractors of the work for certain particulars, from which he favored us with the following account.
>
> The first of September, 1800, the old post road from High St in Newburyport to Charleston Bridge, was surveyed and accurately measured by Mr. John F Somerby and shortly afterwards by General Foster, at the expense of the Beverly and Ipswich Turnpike Company, who made the distance 42 miles and 137 rods. As a proof of the accuracy of the gentlemen, it may be proper to mention that the results of their measurements differed only 10 rods. The Salem Turnpike was shortened the distance 3 miles and 100 rods; so that taking that road, it is only 39 miles and 37 rods, from the head of State Street in Newburyport to the Charleston Bridge; from which (the bridge included) it is 308 rods to the Old State House by way of Prince, Back and Union streets and up Campbell, making the sum total 40 miles and 115 rods.
>
> The Newburyport Turnpike road as measured by the committee appointed by the Court of Sessions is, from the head of State Street, Newburyport, to the old Malden road at Rams Corner, 26 miles and 294 rods; thence in a direct line to Malden Bridge, is 3 miles and 104 rods; over by Prison Point, and across the river to Barton's Point, to the Old State House, is 2 miles and 88 rods; which makes the whole distance from State Street, Newburyport to the old State House in Boston only 32 miles 156 rods; which is 7 miles and

7 New England Reparatory (Newburyport, Massachusetts) July 27, 1803.

8 Newburyport Herald (Newburyport, Massachusetts) February 17, 1804 p3.

9 Political Calendar (Newburyport, Massachusetts) May 3, 1804 p3.

10 New-England Palladium (Boston, Massachusetts January 1, 1805 p2.

11 Massachusetts Spy (Worcester, Massachusetts) January 30, 1805, p3.

269 rods less than the present travel passing over the Salem Turnpike road.

The Newburyport Turnpike road is so nearly in a direct line, that in the first 20 miles all the angles taken together, increase the distance only 83 feet. It is astonishing, that in the first 26 miles, 31 public roads, (and many private ones) are crossed by said turnpike so nearly at right angles, that it is not five rods across any one; and the turnpike road runs nearly central, to the county Essex, and opens a very convenient communication for a large number of the inhabitants, without any expense to them, or emolument to the proprietors.

The proprietors of this road have already expended upwards of 5000 pounds of gunpowder, in blowing rocks, and more than $14,000 in building bridges. Many hills they have reduced from 25 to 30 feet, and raised the valley's as much; not withstanding which, they are determined to make further improvements the next spring, and expect to complete the whole before July next.

It is worthy of observation that this road forms a straight line from the capital of Massachusetts to that of New Hampshire; and such attention has been paid to the smoothness of the surfaces that twelve miles on this, may be traveled with as much ease and dispatch as nine miles on the old road.[12]

[12] Massachusetts Spy (Worcester, Massachusetts) January 30, 1805 p3.

By the middle of 1805, the Turnpike Corporation was starting to sell off some of the road construction equipment. Ox shovels, ploughs, harrows, white rope, cordage, and tools of various kinds were sold at public action.[13] This serves to remind us that the turnpike was constructed without the use of any powered machinery, only the sweat of manual labor and the strength of the hearty oxen. Even though the action was slowing down, another construction worker lost his life when a cave in occurred while he was digging.[14]

In June of 1806, the shareholders had been assessed their 24th assessment and some owed as much as $740 in unpaid prior assessments.[15] Those who did not pay up lost their shares. By October of that year, the 25th and final assessment was called upon completion of the project. Most shareholders by now had invested upwards of $500 per share![16]

Turnpikes were not always the safest way to travel. In 1807, a gentleman by the name of Kidder had his pocket-book, containing $350, stolen while traveling on the Newburyport Turnpike. The robber was dressed like a sailor and was armed with pistols.[17] Robbery was not a frequent news item, but ten years later-"A robbery took place on the Newburyport turnpike on Thursday, beyond the Topsfield Hotel. Two travelers who had hired a man and

[13] Newburyport Herald (Newburyport, Massachusetts) May 24, 1805 p3..

[14] Ibid August 2, 1805 p3.

[15] Ibid June 10, 1806 p1.

[16] Independent Chronicle (Boston, Massachusetts) October 8, 1806 p1.

[17] Newburyport Herald (Newburyport, Massachusetts) March 27, 1807 p1.

his wagon to transport them to Danvers, finding he had money in his pocket, seized him, knocked him down, robbed and tied him. The thieves having made off with the booty, the man gave the alarm. One of them was overtaken and the posse was in pursuit of the other at the last account."[18]

Toll roads were becoming more acceptable, but there was always someone who felt he should not have to pay to use a road even if it did save him time and effort. In 1823, the toll house in Newbury was consumed by fire. James Prince, president of the corporation, offered a $100 reward for information leading to the arrest and conviction of the arsonist.[19]

1823 provided a precursor for the fate of the turnpike. Hotels which had been built in 1804, at great expense, were auctioned off, the one at Lynnfield for $2550 and the one at Topsfield for $3135. A year later, the Topsfield Hotel reopened with the new owner, Cyrus Cummings, claiming: "This hotel is situated on a delightful and commanding eminence on the Newburyport Turnpike about 9 miles from Salem, it is three stories high, has large and commodious rooms, and offers a most agreeable and healthy summer residence to invalids and others who are disposed to spend the summer months in the country. The bar is always supplied with the choicest liquors and the table with the best provisions of the season. Parties of pleasure and others may depend on respectful attendance, and every effort will be used to give perfect satisfaction to those who favor the establishment with their patronage."[20]

1824 brought another kind of business to the turnpike. At the entrance to the Town of Salem and on the Newburyport Turnpike, houses of prostitution were set up. They remained in use for three years until the Court of Common Pleas, sitting at Newburyport, finally convicted John Mumford, even questioning the legality of his claim of ownership of the property.[21]

By 1830, at a stock sale in Boston, shares of stock in the Newburyport Turnpike Corporation were selling for just $7 per share. It was believed that they had cost nearly $500 per share. The rail road was coming soon and the market understood the consequences.[22] In 1835. the Eastern Railroad had surveyed the route from Boston to the New Hampshire line and even compared their route to that of the Turnpike stating that the distance by rail through a straight and level path was only one mile farther than the distance by the Turnpike.[23]

The Topsfield Hotel made the news again in 1841. It was back on the market for sale or to be let. Having been a public house, primarily for travelers along the Newburyport Turnpike, it became apparent that once the railroad took over there would be no need for travelers to seek refreshment or rest along their route to New Hampshire or Maine.

18 Boston Recorder (Boston, Massachusetts July 15, 1817 p125.

19 Newburyport Herald (Newburyport, Massachusetts) July 8, 1823 p4.

20 Salem Gazette (Salem, Massachusetts) July 9, 1824 p3.

21 Essex Register (Salem, Massachusetts) September 24, 1827 p3.

22 Salem Gazette (Salem Massachusetts) September 3, 1830.

23 Gloucester Telegraph (Gloucester, Massachusetts) September 12, 1835 p2..

An obituary for the turnpike appeared in print in early 1842:

> To the Honorable Senate and House of Representatives in General Court assembled.
>
> Your petitioners, proprietors of the Newburyport Turnpike Corporation, beg leave to represent to your Honorable Body, that in the year 1803, a charter was granted to this corporation, as will appear on your records. The proprietors then were desirous of making a road to the capitol as direct as practicable for the purpose of facilitating the intercourse with Boston. To accomplish this expensive object, they were obliged to purchase in fee simple and for the use of the corporation from 70 to 80 acres of land which cost from $20,000 to $25,000, for the purpose of carrying out their project: they also kept the direct line of road from Newburyport to Boston as straight as practicable, which increased the expense far beyond any estimate made so that the road cost $420,000, divided into 995 shares, averaging $420.00 each. On opening the road, it was found that the income was very small and the stock shares rapidly declined. The proprietors have hitherto sustained the road and in addition, have received small, very small, dividends, amounting to the annual income of $1.73 on the share from its commencement. For a few years, previous to the completion of the railroad, the income was slowly advancing and the shares were worth 20 to 30 dollars. When the railroad came into operation, the travelling was diminished more than half and now the proprietors having disposed of all of their available means, except one toll house. They are no longer able to sustain the road.
>
> The turnpike has become necessary for the use of the towns, through which it passes; and many improvements having been made to it by individuals for their own benefit, it would be a serious injury to discontinue the road, leaving the proprietors of the ground over which it passes to occupy and improve all their own right, reserved by the charter, the corporation having a large portion of lands in fee simple.
>
> The proprietors who can no longer keep the road in repair from their income, now beg the assistance of your Honorable Body to relieve the proprietors and to continue to the public, the benefit of the road; which has become so necessary by a long use, by granting the proprietors some small indemnity for the immense loss they have sustained and the right they still retain. They would also respectively represent that many widows and orphans are interested in the road and are suffers; that the many roads laid out by the county commissioners have diverted the travel from the tollgates and lessened the income, and that the railroad has ruined it. They would therefore beg permission of your Honorable Body to suggest the expediency of enacting a law, similar to one existing in New Hampshire, relating to a turnpike

> corporation, or of authorizing the county commissioners to lay out the turnpike road as a common highway to the different towns in which it is located; granting the proprietors such an indemnity as in their judgment may be equitable, on the relinquishing of all claims to the road, by its being discontinued and appropriated to the use of the public. All of which your petitioners would submit to the wisdom and equity of your Honorable Body and as in duty bound ever pray. In behalf of and in pursuance of a vote of the proprietors. Joseph S. Pike, President and R. Stone, Treasurer.[24]

Proving that justice can prevail, on the morning of February 7, 1843, due to the accumulation of snow and ice, the engine of the morning "up train" ran off the track at a location near the Newburyport Turnpike.[25]

By 1849, it was announced that Essex County, Massachusetts had declared that the portion of the Turnpike lying within its boundaries would be made into a public highway. It was understood that the Turnpike lying in Middlesex County would be handled in the same way.[26]

Crossing New Hampshire on the Country Road

> From the earliest settlement of the country, until the present time, the road leading from Newburyport to Portsmouth, passing through Hampton Falls and known as the 'Country Road,' has been an important and much used thoroughfare. Many distinguished persons have passed over it. On December 13, 1774, Paul Revere made his historic ride, traveling from Boston to Portsmouth to inform the Committee of Safety of the British order that no military stores should be exported to America. The reception of this news caused Major John Langdon and John Sullivan to secure and remove the stores in Fort William and Mary at New Castle. The ammunition thus obtained did good service to the colonist at the battle of Bunker Hill.
>
> President Washington rode over this road in his carriage from Newburyport to Portsmouth on October 31, 1789. James Monroe, president of the United States, passed over this road on his northern visit in June of 1817. The Marquis De Lafayette went over this road in passing, from Newburyport to Portsmouth on September 1, 1824. He did not pass over what is now known as the Lafayette road, or stop at the Lafayette House in North Hampton, as many have been led to suppose, as neither were in existence until after that time. General Winfield Scott passed over this road in the spring of 1839, and paused for a time at the tavern in Hampton Falls. He was on his way to Maine to effect, if possible, some settlement of what was known as the 'Aroostook War.' This dispute over the boundary line between this country and New Brunswick was afterward adjusted by the Treaty of 1842.[27]

24 Newburyport Herald (Newburyport, Massachusetts) February 1, 1842 p3.

25 Ibid February 9, 1843 p2.

26 Salem Register (Salem, Massachusetts) May 14, 1849.

27 Warren Brown, *History of the Town of Hampton Falls, New Hampshire.*

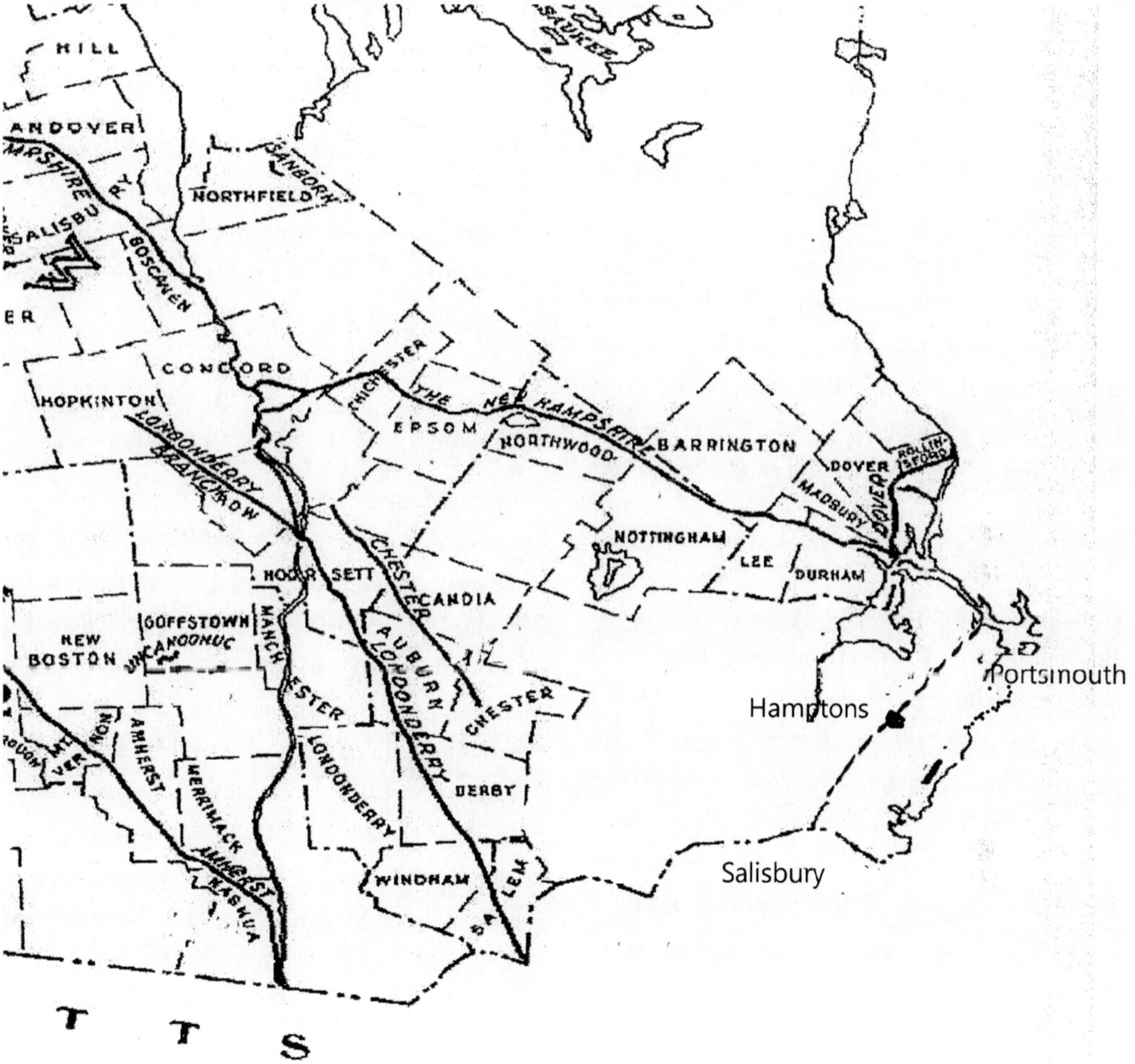

New Hampshire Turnpikes and the Road from Salisbury to Portsmouth[28]

Near the coast, New Hampshire is bounded by a line somewhat north of the Merrimack River on the south and the Piscatagua River on the north and east. To cross New Hampshire from Massachusetts, heading for Maine, one first needed to bridge the Merrimack, then complete the road through Salisbury to the New Hampshire line, pass through the Hamptons and Seabrook, and then bridge the Piscatagua at Portsmouth. For this section of the road there were two jurisdictions involved, Massachusetts and New Hampshire.

A bridge from Newburyport to Salisbury across the Merrimack river had been in use since 1793. An architect from Newburyport, Timothy Palmer, had designed the bridge which

[28] By F.J. Wood (Scanned from *The Turnpikes of New England*) [Public domain].

replaced numerous rope ferries. Amazingly, it was Palmer's first attempt at a bridge design. Prior to this accomplishment, he had been a successful millwright and house wright. His challenge was to build a bridge that would be over 1030 feet long with a span of 160 feet over water that was 34 feet deep. The bridge extended from Newburyport to Deer Island and from there on to Salisbury, and was based on a design by Palladio from the 16th century. The bridge was just east of Interstate 95's current bridge and its original location can be seen from that highway.[29]

Palmer also designed a bridge crossing the neck of water leading from the Piscatagua to Great Bay connecting Durham and Newington. This was part of the New Hampshire Turnpike which ran from Portsmouth to Concord. Being the first Turnpike in New Hampshire, its function was to connect the State Capital with the state's only port. Completion of the bridge in 1794 permitted travel from Portland, Maine to Dover, and then on to Portsmouth, New Hampshire. It seems that no one had dared to try to bridge the wider Piscatagua at Portsmouth until a pile bridge was constructed in 1822. This bridge was located six miles downstream from the original Piscatagua Bridge at Durham.[30]

Prior to the construction of the pile bridge, crossing the Piscatagua at Portsmouth was only accomplished by Rice's Ferry. "In 1692, Mr. John Woodman was authorized to keep a ferry from Kittery to Strawberry by the Court and the town voted their approbation of the same. This was the same ferry kept later by Alexander Rice. Mr. Woodman was the maternal Grandfather of Major Rice."[31]

That original bridge was converted to a railroad and vehicle bridge in 1842, and by 1923 was closed to all but railroad traffic. In 1936, a Norwegian freighter ran into the bridge damaging twenty pilings. Although the damage was supposedly repaired, a local passenger train attempting to cross in 1939 plunged into the river, killing the engineer and fireman.

Following right on the heels of the Newburyport Turnpike, William Bartlet and Nicolas Pike, both of Newburyport, submitted a petition to the Senate and House of Representatives of the Commonwealth of Massachusetts. Having learned that New Hampshire was planning to straighten and flatten the road from Portsmouth south to the line of division between the two states, they realized that a bridge over the Merrimack, located at the center of Newburyport, and at the terminus of the Newburyport Turnpike, coupled with a road through Salisbury, were the only remaining pieces of the puzzle. They also noted that the current post road was sandy and less than desirable.

The petitioners were, of course, asking to be incorporated into a Turnpike Corporation, just as the Newburyport Turnpike proprietors, which also included William Bartlet, had done.[32] In fact, Bartlet had petitioned for a third turnpike by 1805. This one was to extend from the Merrimack Bridge to the state line of New Hampshire, in a direction to connect with the

[29] Frank Griggs, Jr, *Newburyport Bridge*. Retrieved July 17, 2017, from StructureMag.org website: <http://www.structuremag.org/?p=117>

[30] New Hampshire Gazette (Portsmouth New Hampshire) December 9, 1794 p3.

[31] J. L. M. Willis, *Old Eliot*.

[32] Newburyport Herald (Newburyport, Massachusetts) March 19, 1802 p3.

Turnpike to Concord.[33] It is not clear how the Legislators dealt with the requests which followed the Newburyport Turnpike petition. Apparently, the latter two requests were not approved.

It was not until 1825 that a truly suitable route was surveyed and construction begun on a road from Portsmouth to Newburyport.

The New Road from Portsmouth to Newburyport

This important road, which was laid out this last year, has been made in part, and so far as it runs in Portsmouth and Rye, will very soon be passable. When the whole route from Portsmouth to Newburyport shall be completed, it will shorten the distance four or five miles, and afford of the public a much better road than the old one. The whole expense to Portsmouth will probably exceed $5000. The expense to Rye will be somewhat less, it is hoped that a new and straight road will soon be made leading from the Portsmouth Bridge towards Portland. A great road, well made on the seaboard, would be of much public utility. Every possible facility should be afforded for the public travel. And the general disposition to make improvements in roads, augurs well for the future prosperity of the country.

We learn with some surprise, that the towns of North Hampton, Hampton and Hampton Falls have not as yet made any part of the new road! What can be the cause of such negligence? It is very certain that some of the citizens of North Hampton were in the first instance strongly in favor of the new road. We do not know that they have changed their opinions, but why this neglect of duty on the part of three towns? Do they mean to set the laws and the Court at defiance? After several years litigation, a very respectable committee laid out the road, and their report was accepted and confirmed by the Court. A sufficient time has elapsed for those towns to make their part of the road.

First Bridge Across the Merrimack at Newburyport[34]

[33] Ibid March 29, 1805 p1.

[34] *The Massachusetts Magazine* May 1793. Published by Ebenezer Turrell Andrews and Isaiah Thomas

Pile Toll Bridge from Portsmouth to Kittery, Maine[35]

Do they mean to suffer the towns to be indicted? Will it not be much cheaper for the towns to make the road in their own way, than to have it done by virtue of a compulsory process from the Superior Court?

The character of those towns stands too high for us to believe for a moment that they have combined to defeat the making of the road, more especially after two towns have performed their duty, and expended so many thousand dollars.

The new road avoids the causeway in Hampton, where every traveler now has to pay unreasonable toll. This toll has been a source of gain to some of the inhabitants of that town, but it has long been a subject of complaint. Should the citizens of Hampton expect to defeat the new road, in order to reap the benefits of the causeway tolls, they may be disappointed in their expectations. Many towns in the state have to bear greater burdens than Hampton in keeping bridges and causeways in repair, without demanding of every traveler nine pence or 20¢. Every good reason urges that the remainder of the road should be made, and that without any delay. The public interest is paramount to private considerations. A TIMELY HINT.[36]

By the middle of 1826, a new bridge across the Merrimack was underway beginning from the foot of Strong St in Newburyport and ending up

[35] Image Courtesy Portsmouth Athenaeum.

[36] New Hampshire Gazette (Portsmouth, New Hampshire) September 13, 1825 p3.

Toll Bridge from Portsmouth to Kittery with Toll House from Kittery Side[37]

in Salisbury. Upon completion, in October of 1827, it was 1000 feet in length with a draw bridge on the Newburyport side. The total cost was about $66,000 which included the road on to Portsmouth. That road was even covered with two layers of gravel! Now there was a decent road running from Boston to Portsmouth, and it even crossed the Piscatagua River. The next step was to continue that quality of layout and construction all the way to Portland.

Building Roads in Maine

Maine was settled by seafaring men. Hayden V. Anderson, in his book *Canals and Inland Waterways of Maine*, tells us that: "Until well into the 1800's the easiest travel in Maine was by water. Early white settlers came by sea and made homes on the sheltered shores of bays. Later they pushed in small boats up the tidewater rivers to begin the second tier of towns; thus, did John Phinney and his son ascend the Presumpscot in 1735 to found Gorhamtown. Townships such as New Marblehead (Windham) were planned by following a river: 'We began at a Place called

[37] Courtesy of Strawberry Banke Museum, The Garland Patch Photograph Collection.

Sacaripy [Sacarappa] Falls in the Presumpscot River and so as the River runs to Great Sebago Pond.' He then goes on to say:

> Land travel throughout the District was by roads passable only on horseback, the usual mode of travel 'even for some years after the Revolution.' When John Adams came to Maine in the 1770's to visit Pownalborough [Wiscasset] and other court houses as circuit judge, he described part of a trip to Falmouth. 'Riding alone through Saco after dark, he found many sharp, steep hills, many rocks, many deep ruts, and not a footstep of man except in the road; it was vastly disagreeable.' At the time, there was a road from Kittery up the coast to Falmouth, and from Falmouth to North Yarmouth. A road linked Bath and Brunswick, but there was only a spotted trail up the Kennebec. Similar trails were all that connected most of the inland tier of townships in York and Cumberland counties—Berwick, Lebanon, Sanford, Buxton, Gorham, Windham, Gray, and New Gloucester—with the coastal settlements. The settlers of Sylvester's Town [Turner] could reach the nearest important settlement, New Gloucester, only along a spotted trail by way of Bakerstown [Poland]. 'Often a pioneer shouldered a bag of corn and walked twelve miles over this obscure trail to have corn ground at the mill in New Gloucester.'

Maine needed roads, but roads took time and resources to build and maintain. Rivers were always there and could be used in the warm seasons by boats, and in the winter by sleds. Except for periods of flood or drought, they needed no maintenance.

Anderson then comments:

> Road improvement was stimulated by the introduction of wheeled vehicles. The chaise, a one-horse, two-wheeled, covered carriage, reached Maine about 1760, but was used only in town, on special occasions. In 1787 post rider, Joseph Barnard, drove Portland's first regular mail coach from Portsmouth to Portland on the Boston Post Road. The coach, drawn by two horses, was licensed by Congress to carry passengers. The urgent communication needs incident to the Revolutionary War spurred travel and transport developments. 'Mail routes were extended and improved,' and with the demands for passenger-carrying there came a 'concerted movement to develop inland routes suitable for passage of horse drawn vehicles.' Throughout the years before 1790, even as roads proliferated, waterways remained for settlers the usual means of travel and trade.[38]

The only problem with using waterways for transporting goods or people is that, as with trains and airplanes, they do not always go to a preferred final destination. To get where you really wanted to go, roads, and lots of them, were required.

Turnpikes were becoming popular from 1790 to the beginning of the War of 1812. Turnpikes at that time were built by individuals or small corporations to provide suitable roads for

[38] Hayden V. Anderson, *Canals and Inland Waterways of Maine.*

wheeled transportation of goods and people which were used after paying a toll. Towns had neither the wealth nor interest to provide travel outside their own limited jurisdictions. Every part of New England, except Maine, acquired an extensive turnpike network. Particularly missing in Maine, were good roads which linked the coastal towns to the growing inland settlements. Roger N. Parks, in his book *Roads and Travel in New England 1790-1840*, published by Old Sturbridge Village, provides us with further insight into the ways roads were built and maintained in those days.

> Labor on the roads was supervised by surveyors of highways, twenty to thirty of whom were elected annually by some small towns. When John Adams reached the age of twenty-one, he was surprised to find himself nominated for this job and objected that he knew nothing about road making. He was told that 'they make it a rule to compel every man to serve either as Constable or Surveyor, or to pay a fine.' Adams performed his duties in a conscientious way, even though a bridge built under his direction soon was washed away. It was charged, however, that most highway surveyors lacked interest as well as skill and that many of them were concerned only to have the roads past their own houses repaired.

Although manual labor was used extensively for road construction and repair, scrapers drawn by oxen played their part in reducing the human drudgery.

> The Revolutionary War was another factor in the growth of transportation. It created a temporary need for the movement of men and supplies over long distances, but more importantly, perhaps, many a New Englander returned from service with a greater realization that the world was not bounded by the limits of his own town and that distances could be overcome.[39]

From Portsmouth to Portland

Certainly, what was long known as the oldest continually traveled road in Maine deserves mention. That road ran on the shore of the Piscatagua in Eliot, Maine. Willis tells us:

> First, it was the Indian Trail along the banks of the winsome Piscatagua; the river that not only was the attraction of the original Indians, but also as early as 1602, years before Plymouth Rock became the stepping stone, there came to our Eliot and Kittery shores, the companies of English fishermen, who took such pleasure in their successes, that log cabins were built on the very edges of the trail; and these cabins on the rims of both the river and the trail, became their homes year after year. The Indian trail of that early date, came from the old fields of Newichawannock and extended to Kittery Point.
>
> It is not strange that this Indian trail, with its natural beauty, our broad river on one side, and the hills and trees on the other, became not only the

[39] Roger N. Parks, *Roads and Travel in New England.*

> footpath of the Indians, but when the whites came hither for a home, it became a permanent road, never surrendered and never limited; but for the entire three centuries since those early white men built their rude homes for the summer fishing, has it been an open and free road, and the homes of successive generations.
>
> Yes, it is the same old road of comfort and beauty today, it runs on the edge of the Eliot shore, and winding inland for a little, it again reaches out to the riverbank, until the limits of the old town are reached.
>
> The length of the period that the old road has been in open use and travel, we can only judge by the time of the settlement of this region. It was a trail when our settlers first landed; in later years of allusions to it have given it the one name, The Old Road.[40]

A Maine State Highway Commission pamphlet entitled *A History of Maine Roads* describes the early roads in Maine:

> With a ready-made system of transportation all around him, the early Mainer was not about to spend his energies building roads he didn't need. He followed the Indian trails along the banks of rivers and streams making them a little wider and a little deeper. He felled timber and cleared a rough road through the forest to early shipyards. But he had enough to do around the place without yearning to visit the next settlement. Likely as not the next settlement was up river or down and he went by boat. When not used for any period of time most of these primitive roads faded back into the pucker brush.
>
> Roads that did develop pushed their way north from the southern gateway to the region at Kittery [Old Ketterie] from the governing mother colony of Massachusetts. There is an early report of the Massachusetts commissioners riding into the district and getting no farther than Wells because of the condition of the. roads. Thereupon they ordered the inhabitants of the Saco, Wells, and Cape Porpoise area to 'make sufficient roads within their town so that travelers could move from house to house' or suffer a fine - an edict that probably affected the people in that area like other pronouncements in the Crown's name.
>
> As a result of the discomfort of the commissioners, however, stretches of a shore road were built along the Maine coast from Kittery extending northerly to Portland [then called Falmouth.] The first section of this early road was laid out in 1653 and called the 'Kennebunk Road by the Sea,' It was staked out to a definite width but nothing was done to improve it as a highway. Although bearing the euphonious name of the King's Highway it consisted of nothing more than two cart wheel ruts with a horse path in the middle. In his novel *Arundel*, Kenneth Roberts describes this early ancestor of Route 1 and 1-A as a series of bog holes where the horse could slip into mud up to its

40 J. L. M. Willis, *Old Eliot.*

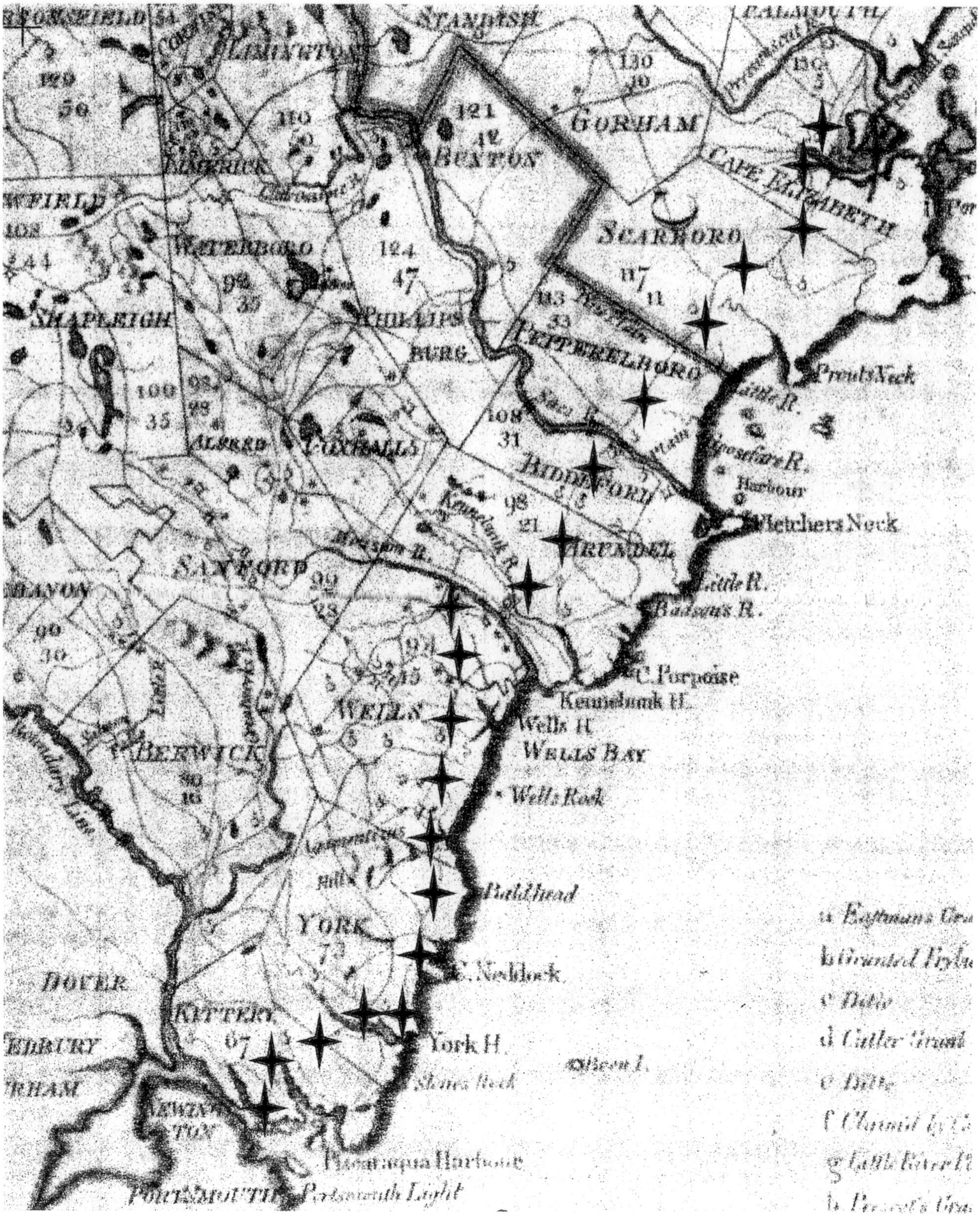

Carleton's 1802 Map Showing the King's Highway from Kittery to Portland

withers. There were no bridges and waterways along the coastline were forded where possible at low tide and crossed with a ferry as travel grew in volume.

This first King's Highway was the only means of communication between the Province of Maine and Massachusetts for more than a century. In the middle of the 18th century an upper- King's Road was laid out connecting the thriving villages of York, Wells, Kennebunk and Saco. When Benjamin Franklin was appointed Colonial Postmaster, he established a weekly mail service east of Boston thereby giving the name of 'Post Road' to the newly completed upper King's Highway. Large granite blocks were placed along the post roads radiating outward from Boston. In Maine, the milestones were set along the early Post Road from Kittery to Portland in 1760. Present day U.S. Route 1 and 1-A have absorbed sections of the old Post Road but here and there where the old road is still on its own, the original name. 'King's Highway' is displayed - a genuine relic of the past.[41]

As more New Englanders became directly affected by the poor condition of their roads, enterprising men with money to invest moved in to fill the need. Toll bridges and toll roads became very popular in most of New England. Maine was an exception. As an example of what could have been, and later came to be, we have the original Maine Turnpike.

On August 6, 1802 a bill was presented to the Senate and House of Representatives of the Commonwealth of Massachusetts in General Court assembled:

> The subscribers humbly show that in their opinion the interest of the commonwealth, would in several respects be greatly promoted by an improvement of the roads in the Eastern part of the same. That there is no better way of effecting this improvement, than by the establishment of turnpikes. They beg leave therefore to propose for the consideration of the Court, the expediency of making legislative provision for the purpose of incorporating the company to make a direct road from Piscataway River to Portland, thence to Augusta and from Augusta to the town of Prospect on the Penobscot River. It will be an expensive undertaking; but with the aid of government and the exertions of individuals it may be done in a short time; and the benefit will doubtless compensate for the expense. Your petitioners therefore humbly pray that your Honors would make a grant to your petitioners and others, who may be admitted to join them, of ten townships of the undisposed lands belonging to the gov't, in the district of Maine, or such other number as your honors shall judge best in aid of the undertaking and incorporating.[42]

D. L. Soucy, in a *Maine History News* article, talks about this proposed turnpike.

[41] State Highway Commission, *A History of Maine Roads 1600-1970.*

[42] Kennebec Gazette (Augusta, Maine) August 6, 1802, p1.

The Maine Turnpike Authority we know of today is not the first cadre of highwaymen authorized by judicial means in the state. Way back before we were known as the state of Maine, our trustees by way of the Massachusetts legislature of the time granted an incorporation known as the Maine Turnpike Association. Its purpose was to build and maintain a roadway, paid for by tolls of course, that would stretch from the New Hampshire border in Kittery, thence to Portland and then on to Augusta, in as straight a line as possible. The act was approved in January of 1803, and in spite of much chest thumping and bragging by the proprietors of the corporation, the little bird never left the nest and not one shovel of earth was turned over to build the turnpike.

In spite of the lack of completion of the project, the concept of a roadway to what would become our state capitol still managed to pervade the thoughts of planners along the seacoast communities. If you pay close attention to roadmaps that have been compiled over the years, you can see that what now is known as US Route 1 travels over a fairly straight route until it reaches the mid-coast area and veers away from our capitol to follow the coastline to Calais. That same route is very nearly the route suggested way back in 1802.

According to the legislative record of this act forming the corporation, the following lays out what the turnpike is supposed to look like:

'... shall be a corporation by the name and style of The Maine Turnpike Association, with all the powers and privileges usually given and belonging to similar corporations, for the purpose of laying out, making & keeping in repair a turnpike road from the line of the State of New Hampshire to Portland, and from thence to Augusta Bridge, upon as straight a line as circumstances will admit; and erecting and keeping in repair such bridge or bridges as may be necessary on said route, which turnpike road shall not be less than four rods wide, and the part to be travelled on not less than twenty four feet in width, in any part thereof; and when said Road or any ten miles thereof-shall be sufficiently made ... then the said Turnpike Corporation shall be authorized to erect turnpike gates on the said road, at such place or places as the said committee, of the said Court of Sessions, and the said corporation shall judge necessary and convenient, for collecting the toll, provided that no turnpike gate be erected on, or any toll demanded on any part of the present travelled Roads; the said gates to be not less than ten miles distant from each other ...

And of course, tolls were allowed to be collected and the rates approved by the legislature:

... shall be entitled to receive of each traveler or passenger at each of the said gates, the following rates of toll, For each coach, phaeton, chariot or other four wheel carriage, drawn by two horses twenty five cents; and if drawn by more than two horses, an

> additional sum of four cents for each horse; for every cart or wagon drawn by two horses or oxen ten cents, and if drawn by more than two horses or oxen, an additional sum of three cents for each horse or ox, for every curricle fifteen cents; for every chaise, chair or other carriage drawn by one horse twelve cents; for every man and horse six cents; for every sled or sleigh drawn by two oxen or horses nine cents, and if drawn by more than two oxen or horses, an additional sum of two cents for each ox or horse; for every sled or sleigh drawn by one horse eight cents; for all horses, mules, oxen, or neat cattle, led or driven, besides those in teams or carriages one cent each; for all sheep or swine at the rate of six cents, for one dozen ... [43]

Before construction could begin, they first needed to survey the route from Portsmouth to Portland. In October of 1802 it was noted:

> The petitioners for a turnpike road from Piscataway River to this town and from this town to Augusta, have employed a gentleman of character and abilities to take the courses and distances of the present road from hence to the great bridge over that river, and to mark out the nearest route on which a turnpike can be made. He is now upon that business. The committee has been lately engaged in exploring the nearest route from this town to Augusta. The undertaking is important, and if affected must prove eminently advantageous to the public. It is therefore hoped that the undertakers will meet the success in their application, and be encouraged to pursue the object of that, until they reap the fruits of their exertions; and the good citizens of this Commonwealth enjoy the pleasure of traveling on these routes with ease and expedition.[44]

A majority of the members of the "Maine Turnpike Association," held their first meeting April 1, 1803 at the home of Thomas Peck to establish officers, rules, and regulations for the corporation. The subscribers were: Stephen Waite, Peter Warren, John Deane, Joseph H. Ingraham, James Neal, Stephen Patten, Samuel Freeman, Woodbury Storer, Isaac Parker, Daniel Tucker, Robert Boyd, Willian Symmes, and Isaac Ilsle.

By March of 1806, the Maine Turnpike Corporation was already feeling pressure from possible competitors. On the 14th of that month, a petition was presented to the House of Representatives for the Commonwealth of Massachusetts requesting the incorporation of a company to build a bridge over the Fore River between Portland and Malam's Neck in Cape Elizabeth. Concurrently, the same petitioners asked that the Maine Turnpike Association's should have that portion of their route repealed.[45]

43 D. L. Soucy, Maine History News, *The Maine Turnpike (that never was.)* June 23,2012. Retrieved July 17, 2017, from Touring-MainesHistory.wordpress.com website <https://touring-maineshistory.wordpress.com/2012/06/23/the-maine-turnpike-that-never-was/>.

44 Gazette (Portland, Maine) October 11, 1802 p3.

45 Eastern Argus (Portland, Maine) March 14, 1806 p3.

A couple of months later, a meeting was called by the proprietors of the Maine Turnpike Corporation, to discuss the idea of relinquishing that portion of the proposed turnpike from the southwest end of Vaughn's Bridge to the southwesterly banks of the Saco River to the proprietors of Vaughn's Bridge, or to merge the two entities.[46]

Vaughan's Bridge Corporation was formed by an Act of the Massachusetts Legislature in 1794. The bridge extended from Bramhall's Point in Portland and over the Fore River to Jacob Brown's farm, in what was then Cape Elizabeth. Later on, that landing point was Main Street, Route 1, in South Portland.[47]

Prior to Vaughan's Bridge, travelers coming from the south went on Westbrook Street into Portland over the bridge at Stroudwater. That route was eliminated when the airport was built.[48]

Getting the last small stretch into Portland would become the Achilles Heel for the project. The final gasp would come in April of 1809, in a further petition to the Senate and House of Representatives for the Commonwealth of Massachusetts:

> Humbly show, the subscribers, directors of the Maine Turnpike Association, that by an act of the legislature passed March 8 1803, the said Maine Turnpike Association with vested with the right and power to erect and make a turnpike road from the line of New Hampshire to Portland, and the further right to erect and keep in repair such bridge and bridges as might be necessary on said route. That said corporation has laid out and located the part of the route from Portland to the line of New Hampshire, but before that part of the route so located can be fully completed a bridge must be erected from the bank of Fore River in Portland, over the said Fore River to Malam's Neck (so called) in Cape Elizabeth and the expense of building said bridge will be so great, that no encouragement at present under the existing act is offered for erecting said bridge, as the toll which the corporation might receive now, provided said Bridge was erected, would not be sufficient to keep the bridge in repair and of course be no remuneration for the large sums of money which must necessarily be expended to complete it. Your petitioners further beg leave to state to the Honorable Legislature, that repeated applications have been made by various persons to complete the most difficult parts of the route, for the toll now granted by the act of incorporation, but owing to the magnitude of the undertaking of building said bridge, and for merely the consideration of receiving such toll as it is now granted, this part of the route is unfinished and the residue of the same remains also unfinished. Your petitioners therefore, humbly pray, that your honors, (if on due consideration it should be thought expedient) would pass an act authorizing the Maine Turnpike

[46] Gazette (Portland, Maine) May 5, 1806. p3.

[47] John T. Fagan, *The Charter and Ordinances of the City of Portland.*

[48] Maine Memory Network, Item 76191.

> association aforesaid to receive such reasonable toll from those who may pass said Bridge (when it shall be completed) with teams, horses, carriages or otherwise in addition to the toll which they might now receive, were the bridge erected as in their wisdom they may deem equitable. And as in duty bound will ever pray. John Thrasher, Richard Honeywell, Henry F Dupee.[49]

Apparently, no portion of the turnpike was ever completed in the case of the Maine Turnpike Corporation! The road from Portsmouth to Portland could claim one small success in the way of a turnpike. The First Cumberland Turnpike was created as a corporation in 1802 and was completed shortly thereafter. Horatio Southgate was the leading proponent, and he even lived right at the beginning of the new road. He became treasurer and Chairman of the Standing Committee. The Turnpike ran for only one and one-half miles in Scarborough, from Dunstan's Corner towards Portland. The road crossed swamp lands and saved a mile in travel distance over the old Post Road.

By 1834, strong opposition to turnpikes nearly turned the turnpike into a free road, but the corporation prevailed and it continued as a toll road until 1847 when the proprietors of Vaughn's Bridge rented the road for $100 per year. Four years later it returned to its toll road status and was eventually purchased by the county and the town of Scarborough.[50]

In the next chapter, we will learn about Josiah Paine, a stage line operator. Paine became so irritated at the notion of having to pay twenty-five cents toll each time his stages used the First Cumberland Turnpike, that he built a parallel road that is still known today as Paine Road.

[49] Gazette (Portland, Maine) April 3, 1809 p4.

[50] F.J. Wood, *The Turnpikes of New England.*

Chapter Two

Forming the Stage Lines

An Early Stage Line Out of Boston[1]

Yankees have always been known for their keen business sense. Even before the Revolutionary War, men in possession of horses and wagons began considering ways to convey the public from town to town. As early as 1761, passengers were being transported along portions of the road between Portland and Boston.

[1] Courtesy of Strawberry Banke Museum, Portsmouth, New Hampshire.

The first public stagecoach line from Boston (to anywhere) was established in 1761 by Bartholomew Stavers who had his headquarters at the Sign of the Lighthouse in the North End. The starting point was in Charleston, in order to save the trouble of ferrying, and the coach ran to Portsmouth, New Hampshire. His conveyance was called the 'Portsmouth Flying Stagecoach' and was built to carry 'four persons inside each person to

> pay 13 shillings and six pence sterling up to Portsmouth and nine shillings to Newburyport; to set out every Friday morning between 6:00 and 7:00; to put up at inns on the road where good entertainment and attendance will be provided for the passengers in the coach. Returning to leave Portsmouth every Tuesday morning.' The owner also announced that as this was a convenient and genteel way of traveling and much cheaper than hiring carriages or horses, he hopes gentlemen and ladies would encourage the same. This first stage line was regarded with much astonishment and its establishment was looked upon as a great achievement, and it may be imagined that no railway was later greeted with more enthusiasm than was the opening of this first stagecoach line between Boston and Portsmouth. Of the comforts of Mr. Staver's equipage, the least said probably the better, for the coaches of ye olden time have been likened to everything from a distiller's vat to a diving bell, with sundry violoncello cases hung equally balanced between front and back springs in between. The motion has been said to resemble 'a ship rocking or beating against a heavy sea and straining all her timbers with a low moaning sound as she drives over the contending waves.'[2]

Bartholomew Stavers was a fascinating personality, but he was not alone. His brother, John, was a partner in the Earl of Halifax Inn located in Portsmouth. The arrangement lasted for about seven years, at which time Barth took on the job of Post Rider which meant he only needed one horse and could carry no passengers. Barth and John then tried to sell the stage line, but only with the proviso that the new owner would have to carry on the stage business. There was no intention that Barth should actually carry the mail himself, but rather the new stage line owner would be further enticed by the ten-pound sterling amount which would be paid to him annually for also carrying the mail![3] As a footnote, it was mentioned that Barth would continue managing the stage line until a buyer appeared.

In 1766, Barth had brought the news to Portsmouth of the total repeal of the Stamp Act, which had caused the colonist grief for years. Portsmouth rejoiced with bells ringing, guns firing, colors flying, and bonfires. That news was not Barth's cup of tea, for he was a confirmed Royalist![4]

By 1768, the brothers' horses were afflicted by an illness, then called the "mortal distemper." They advertised for replacement horses after having missed one week's trip to Boston.[5]

The stages were still running in 1772 when Stavers and Hart announced that their two lines would both run from Boston to Portsmouth, one leaving Tuesday and the other on Friday. Only a small bundle was allowed each passenger at

2 State Street Trust Company, *Some Events of Boston and Its Neighbors*.

3 New-Hampshire Gazette (Portsmouth, New Hampshire) July 10, 1767.

4 Boston Evening Post (Boston, Massachusetts) May 26, 1766 p3.

5 New-Hampshire Gazette (Portsmouth, New Hampshire) March 4, 1768 p3.

no additional charge. John Stavers and Benjamin Hart signed the advertisement.[6]

In 1774, Bartholomew Stavers had found that "steady adherence to government and opposing the lawless conduct of the Americans" would be his mantra. Staver's had gone out of his way to express unpopular political ideas. He favored obedience to Royal authority and had expressed this position since 1765. His community, on the other hand, questioned the British authority and during the 1770's, Stavers gained a reputation of being obnoxious. He remained outspoken until the summer of 1774, when a group of Portsmouth citizens attacked his house and dragged him out for a beating. Stavers was not killed by the mob, nor did they tar and feather him. His furniture was smashed but some of the better people in Portsmouth put a stop to the disturbance. Stavers wasted no time leaving. He was hoping that someday the conditions in America would show a better aspect.. His intention was for this to only be a temporary retreat, so his wife and children stayed behind. Sadly, his exile became permanent.[7]

Nine years after leaving Portsmouth, Bartholomew Stavers died in Castle Camps, Cambrigeshire, England. He left a will which tells us that he was no pauper, nor just a poor mail carrier, as some have suggested.

Stavers left a sizable estate to his wife Martha, then residing in Portsmouth. His eldest son, Bartholomew Stavers, now at sea, received his copyhold estate in Castle Camps while his youngest son William, born three months after he left Portsmouth, was also left a sizable sum. He also left money to the Church Wardens at Castle Camps for use by the poor and for church maintenance, as was commonly done by landed gentry.

Barth interestingly mentions a son Andrew, who was born before his marriage to Martha, and another illegitimate son, George Stavers Libboo, in Portsmouth.[8]

Brother John, a member of the Masonic Order, stayed in America, living until 1797. At age 83, his death was the result of a carriage accident while he was out riding with one of his daughters.[9] John and his wife, Katherine, had a son born in 1767, also named William, who apparently carried on in the stage business for his father and absent uncle.[10] William inherited much of his father's property in Portsmouth.[11]

Eight years before his father's death, William had become something of a hero in Portsmouth.

Portsmouth, October 10, 1789

Last evening was committed to the jail in this town, one Markus Plantain, a Frenchman of the province of Bayonne, charged with having burglarously broken into the house of the honorable Paine Wingate, Esq. of Stratham, early on Wednesday morning last, and taken from thence a hat, an overcoat, several large silver

6 Boston News-Letter (Boston, Massachusetts) July 30, 1772 p.3.

7 Christopher Moore, *The Loyalists Revolution, Exile, Settlement.*

8 National Archives, UK.

9 New-Hampshire Gazette (Portsmouth, New Hampshire) October 3, 1797.

10 Ancestry.com, Family Tree for John Stavers of Portsmouth, New Hampshire.

11 New Hampshire Will and Probate Records, Rockingham Estate Papers No.6339-6445.

> spoons, etc. He had proceeded as far as Rowley, with his ill-gotten wealth, when he was overtaken by Mr. Staver's in the mail stage, who knowing him by his senatorial habit, invited him to jump into the carriage, it being empty. The bait was immediately swallowed and he was glad to rest his weary limbs, little suspecting that his journey would end so unpleasantly; he was confined as if in jail until the return of the stage, when Mr. Staver's, not willing to leave him with strangers, brought him to this town and delivered him into the hands of justice.[12]

There was a short-lived stage line from Newburyport to Boston, which began in 1773 under the control of Ezra Lunt. Lunt drove the stage himself, in addition to publishing the Essex Journal and Merrimack Packet.[13] It is not clear which was his day job, but he only continued the stage line until 1775, at which time he went off to fight in the Revolutionary War.[14]

In the fall of 1800, William Stavers and Josiah Paine were subscribers to an advertisement for a line of stages running from Boston to Portland. The schedule called for departing Portland every morning, except Sunday, and arriving in Portsmouth by the same evening. Then a stage would leave Portsmouth the next morning, arriving Boston that same evening.

Paine and Stavers also stated that a separate stage would leave Portsmouth on Tuesday, Thursday, and Saturday mornings, arriving in Portland the same evening. The return stage would leave Portland every Monday, Wednesday, and Friday, and arrive Portsmouth the same evening.

They then declared that "The Proprietors will provide good carriages and horses, and careful, obliging drivers." Stage books recording who wanted to travel and when, were kept at Mr. Kendal's Bunch of Grapes in Boston, at Mr. Staver's in Portsmouth, at Mr. Rice's in Kittery, and at Mr. Paine's in Portland. These, of course, were all taverns and the last one mentioned reminds us that as an astute business man, what better arrangement could you have than to convey your tavern clients to the very doorstep of your own tavern. Who was this new stage man on the scene?[15]

In 1792, an inspired 32-year-old mail carrier by the name of Josiah Paine decided that in addition to carrying the mail he could profitably also distribute newspapers along his mail route. At the time, Paine was a resident of Portland so his first route was within Maine extending from Portland to Berwick.[16] By August of 1792 Paine was adding on a newspaper distribution route in southern New Hampshire.[17]

When springtime came to New England in the year 1805, Mr. Paine announced a line of stages running all the way from Boston to Camden in the district of Maine! This was a huge undertaking. He had stages from Portland to Boston, Portland to Wiscasset, Wiscasset to Camden, Brunswick to Augusta, Boston to

12 Essex Journal (Newburyport, Massachusetts) October 14, 1789 p3.

13 Boston Gazette (Boston, Massachusetts) May 17, 1773.

14 Essex Journal (Newburyport, Massachusetts) January 25, 1775.

15 Impartial Register (Salem, Massachusetts) October 2, 1800 p1.

16 Edward Henry Elwell, *The Successful Business Houses of Portland.*

17 Phoenix (Dover New Hampshire) August 29, 1792 p2.

Portsmouth, and Boston to Newburyport. Mr. Paine's name headed the list of propritors. He had certainly come a long way in thirteen years.[18]

A month later, Paine advertised for another stage line which just ran from Portland to Portsmouth over the earlier Piscataqua Bridge, passing through Dover and Berwick. There he added the note that fourteen pounds of luggage would be allowed on the stage at no charge.[19]

A few days later, Mr. Paine advertised that he had a new line of stages operating on the upper road from Portsmouth to Portland and back again. This time he added a few more taverns to his list of where the stage books were kept. These included Samuel Bowles at Portsmouth, Colonel Coggwells at Dover, George Frost at Berwick, and Major Jeffords in Kennebunk, along with his own tavern in Portland.[20]

Typically, a stage line would look to area farmers to supply the hay necessary to feed the large band of horses necessary for pulling the stages. In 1805, Mr. Paine advertised: "A generous price will be given for a few tons of old hay of the first quality."[21] Of course, Paine was always on the lookout for good horse flesh and in 1807, he advertised: "Wanted to purchase six or eight good horses, suitable for stage horses."[22]

By 1810, Mr. Paine was doing better with the stage line than he was with his tavern in Portland. In that year, he offered to lease his tavern. Paine notes in his advertisement that it was a large and convenient house suitable for a tavern or boarding house, and that it had convenient out houses and stables which could accommodate fifty horses.[23] Almost three years later, Josiah Cummings notified the public that he was now running a house of entertainment on the upper corner of Court Street formerly occupied by Josiah Paine.[24] In 1813, Mr. Paine was offering for sale, or exchange for buildings in Portland, his small farm located nine miles from Portland.[25] Six months later Mr. Paine was advising the traveling public that the stage books for Portland were now being kept at the public house run by David Green on Maine Street, formerly the mansion house of Thomas Motley, Esq.[26]

A letter to Major Russel, editor of the Boston Centinel newspaper, explained a "difficulty" experienced by the stage line with respect to two customers who had been found to have brought more luggage than allowed with them on the stage at Kennebunk. Although Mr. Heyward and Mr. Buncker were allowed only 14 pounds of luggage each, it was determined that they had 99 pounds. They refused to pay for the additional luggage but were permitted to proceed to Portland. On arriving there they were asked again to pay the extra charge and again refused, but were allowed to take their

18 Newburyport Herald (Newburyport, Massachusetts) May 7, 1805 p1.

19 United States Oracle (Portsmouth, New Hampshire) October 25, 1800 p4.

20 Ibid (Portsmouth, New Hampshire) November 1, 1800 p4.

21 Eastern Argus (Portland, Maine) August 2, 1805 p4.

22 Gazette (Portland, Maine) May 4, 1807 p3.

23 Ibid (Portland, Maine) September 10, 1810 p4.

24 Ibid (Portland, Maine) May 3, 1813 p4.

25 Ibid (Portland, Maine) November 8, 1813 p4.

26 Eastern Argus (Portland, Maine) March 31, 1814 p3.

luggage anyway. The next day they returned to the stage office and did pay the extra charge which they owed. The stage agent then commented that he hoped these travelers would proceed more peaceably if they made another excursion to the Eastward. What a contrast to today's traveling fiascos. Those living 200 years ago were certainly more trusting and genteel than we are today. The letter was signed by Josiah Paine, now titled stage agent.[27]

In 1817, Josiah Paine advertised his two-story brick house on Free Street available to rent.[28] By 1819, Josiah Paine was included in a list of delegates to form a Constitution for the new State of Maine. He was then living in Buxton. Certainly, he had proved his worth to the community by that time.[29]

Mr. Paine lived until 1825. A mortuary notice gives additional details of his life:

> On Friday evening last, Josiah Paine, Esq. age 65 died. There is no man perhaps in this state, to whom the public are more in debt than to the deceased. He has been a mail contractor for about 30 years and had been constantly improving and enlarging the mail routes in the state, until the decline of his health, which commenced about four years since. It is well known that the mail from Boston to Portland has been conveyed with more rapidity and punctuality for the last 20 years, than (for the same distance) in any part of the United States, and this was owing principally to the energy and perseverance of the deceased. Mr. Paine was liberal in his religious and political opinions, a man of strict integrity, a kind husband, indulgent father, and faithful friend. He has left a widow and 12 children with a large circle of friends to mourn his loss.[30]

It took until 1833 for his executor to complete the sale of Mr. Paine's holdings which included;

> A store in Middle Street,
>
> A house at 1050 Free Street,
>
> Three lots on Free Street,
>
> Mansion-house on Congress Street at the head of King Street,
>
> Two lots on Federal Street,
>
> Six lots on Middle Street,
>
> Two lots on Noble Street,
>
> A lot on Danforth Street,
>
> Quarter part of a lot on Fore Street with flats and wharf,
>
> Five lots on Vaughn Street,
>
> Shares in the stock of the Eastern Stage Company,
>
> Shares in the Portland Stage Company,
>
> Shares in the Bank of Portland.[31]

His total estate was in excess of $27,000 which today would amount to roughly one million dollars. Considering the increased value of Portland real estate, today his total worth would be in the tens of millions of dollars.

27 Ibid (Portland, Maine) September 30, 1817 p1.

28 Gazette (Portland, Maine) October 7, 1817 p3.

29 Bangor Weekly Register (Bangor, Maine) September 30, 1819 p3.

30 Eastern Argus (Portland, Maine) January 24, 1825 p3.

31 Ibid (Portland, Maine) June 14, 1833 p2.

Paine's holdings in the Eastern and Portland Stage Companies leads us nicely into the additional stage lines which traveled on the Boston to Portland route. These two lines were in their heyday at about the time Mr. Paine died.

Another early stage man was Benjamin Hale of Newburyport. Following in his father's footsteps, Benjamin learned stagecoaching and horse handling at an early age. By driving a stage for his father, Jacob Hale, he became an expert in the field.[32]

By 1804, he was advertising a route from Newburyport to Boston, running daily, except Sunday, during the summer.[33] For the winter months he offered, for the several lines of stages, service from Newburyport to Boston and Newburyport to Portsmouth. Basically, there were three stage lines covered by this advertisement. The first was the Mail Stage, the second was the Newburyport Stage, and the third was the Portsmouth Stage. The list of proprietors for these lines included: Josiah Paine, Benjamin Hale, Zachariah Davidson, Stephen Bartlett, and Josiah Patch.[34]

Following his father's death in 1805, Benjamin and his brother Joseph took over the father's stage business, now calling their entity the Hale Stage Line. While Benjamin focused on running and sometimes personally driving the stages, Joseph specialized in coach manufacturing.[35]

Apparently, Benjamin was an inventor as well as an entrepreneur. In April of 1818, he received a patent on a design variation to the coach which allowed greater comfort to the passengers by using the weight of the baggage in the rear boot to counterbalance the weight of the passengers inside the coach.[36]

Prior to 1819, the term "Eastern Stage" was applied to many stage lines running out of Boston. Virtually any stage which traveled north of Boston was called the Eastern Stage. Some of these were the stages running easterly into Maine and New Hampshire. Reading the accounts of these stages in the newspapers, one immediately realizes the strategic importance of stage lines in that era to convey news, particularly of the 1812 war with England. As ships were taken or burned by both the Yankees and the English, news of their fate spread quickly using the medium of stage passengers. These passengers were the daring and inquisitive ones who learned what they might while away from home and could not wait to report their findings on their return.[37]

Benjamin continued to run his stage business until 1818, at which time he joined with other small stage lines to form the Eastern Stage Company.

The Eastern Stage Company was incorporated in 1818, interestingly in New Hampshire, not Massachusetts, and no doubt absorbed many of the smaller stage lines operating north of

[32] Dr. G. William Freeman, *Benjamin Hale*.

[33] Newburyport Herald (Newburyport, Massachusetts) June 5, 1804.

[34] Political Calendar (Newburyport, Massachusetts) January 7, 1805.

[35] Dr. G. William Freeman, *Benjamin Hale*.

[36] Newburyport Herald (Newburyport, Massachusetts) April 10, 1818.

[37] Newburyport Herald 9Newburyport, Massachusetts) July 28, 1812 p3.

Boston.[38] In January of the following year, the clerk was busy collecting old stock shares in the various small stage lines which had been absorbed by the Eastern Stage Company. New shares were issued to the old proprietors.

By 1824, there was already dissention in the ranks as a onetime proprietor of the Eastern Stage Company set up what was described as an "opposition stage line" on the route between Boston and Exeter, New Hampshire.

William Potter set up a new stage running between Boston and Newburyport in opposition to that of the Eastern Stage Company He then extended that service to include a line from Newburyport to Exeter.

Potter claimed that the Eastern Stage Company had then added a new stage destination to its line, also traversing from Newburyport to Exeter. To minimize the impact of its competitor, the Eastern Stage Company offered reduced rates for this line.[39] The fare from Exeter to Newburyport was 50 cents with the rate on to Boston being one dollar. Their reason for doing so was stated as: "They regret the necessity of the step, but have been led to it, from the fact, that the opposition stages had taken passengers from the stages of the company at reduced prices, thereby opening a warfare destructive to the stage interest. The fares in all other stages of the company will remain as heretofore."[40]

[38] Newburyport Herald (Newburyport, Massachusetts) January 19, 1819 p1.

[39] Boston Commercial Gazette (Boston, Massachusetts) January 8, 1824 p3.

[40] Newburyport Herald (Newburyport, Massachusetts) January 9, 1824 p4.

Potter seems to have been the Stage Agent in Salem, Massachusetts, managing the Sun Tavern there. The battle raged on from January through March of that year, Potter on one side and Jeremiah Coleman, Agent for the Eastern Stage Company, on the other. It ended with an attachment of Mr. Potter's stagecoach and a suit against him by the Eastern Stage Company.[41] In those days, states would allow corporations to exercise a monopoly. In the case of stage lines, this would be a monopoly over certain routes.

Potter tried to continue his role as stage line owner, and in October of 1824, he advertised an accommodation stage between Salem and Boston. The mortgage of his tavern was for $10,000, and by November of 1826, his farm in Hamilton, as well as a dwelling house and stable in Salem, were being sold by assignees. William Potter's tavern, The La Fayette Coffee House in Salem, was sold at auction by his creditors in March of 1827. Without a mail contract from the Postal Service, stage lines across the country rarely lasted beyond a few years

In the early days, there were two classes of stagecoaches, mail and accommodation. Once Congress had approved of carrying passengers on a mail stage, which was under strict scheduling requirements as asserted by Congress, passengers found that the mail always came first. This also meant that the stage would need to stop at every little town along the way to deliver and pick up mail at the local post office. Stage lines soon figured out that their human cargo would be better served if they also offered a stage that would cater more to the

[41] Salem Gazette (Salem, Massachusetts) March 12, 1824 p4.

passengers. This was called an accommodation stage and it took a longer time to arrive at its destination, but was more considerate of its passengers

The Eastern Stage Company seemed to be headquartered at Newburyport, Massachusetts, even though the incorporation was under the laws of New Hampshire. The Boston terminal was at the Ann Street House. By 1825, the company owned 35 coaches, 12 chaises, and 287 horses. 25 men were employed at Newburyport alone, to build and maintain the rolling stock. That year they declared an 11% dividend.

The company was a thriving business and in 1829 applied for incorporation in Massachusetts. That charter was granted in March of 1830

In 1833, the company became free of debt while it employed 500 men. They had added interests in hotels, taverns, blacksmith shops, and stables. In April of 1829, the company purchased one half interest in the Merrimac Hotel in Newburyport.[42]

By 1836, the Eastern Railroad was chartered and that spelled the end of the golden era for the Eastern Stage Company. The railroad and competitive stage lines forced the company to reduce fares and that was a disaster.

Another case of competition occurred in 1837 with a similar result to the previous instance.

To the Public

It is probably well known to the travelling community, that since September last, there has been a stage running tri weekly, from the Farmers Hotel in Portsmouth, through Greenland, Stratham, Exeter, Kingston, Haverhill to Andover, and from thence to Boston by the railroad and that since July last the stage has run daily on a new arrangement. The proprietors of this line of stages take great pleasure in saying, that thus far the public have liberally sustained this enterprise. Within a few days, however, there has been a manifest disposition on the part of the old 'Eastern Stage Company' to drive this line from the road, and with a view to this purpose, the said company has resorted to their old trick of reducing the fair. But will the public be deceived by this paltry contrivance? How long the Eastern Stage Company will keep their fare at the present rates, any Yankee at all acquainted with its history, can easily guess. I ween it would be just so long as it will take them to drive all competition from the road and no longer. So soon as the road is clear, they would immediately resume their old rates. This was the game they played last winter, and they have repeatedly done the same before. No sooner is competition put down, then up comes the $3.50 fare again and if perchance the traveler has more than one trunk or bundle he is charged for them in addition.

Will not the public, then, encourage this competition, by which means only

[42] Trumpet and Universalist Magazine (Boston, Massachusetts) April 25,1829 p171.

> will they be saved from the grinding system of an overgrown monopoly? Is it not better that the fare to Boston should be permanently fixed it $2.75 (which will afford but a fair profit,) then at $3.50 or temporarily fixed at $2.00 to be increased $1.50 as soon as the whole business of the road is monopolized by this company as formerly? I trust the travelling public will act with discrimination in this matter, their good sense must certainly lead them to guard against unreasonable charges, by encouraging competition. The staging business between this place and Boston has for many years been engrossed by this overgrown company. Although attempts at competition have, at times, been made, the great wealth of this company, accumulated from their enormous profits, has crushed them almost in embryo. The same system of under carrying was resorted to; and in consequence of this greater wealth, it was enabled to hold out the longest and run down all competition.
>
> But in the present enterprise the proprietors of the new line have so far, (it being now nearly a year since they commenced running,) been successful and for this they are grateful. It remains now for the public to sustain them against the losing system.[43]

As a further injury, the State of New Hampshire refused to extend the company's charter in 1837.[44] In February 1838, directors voted to sell all real estate and business property, and bring its affairs to a close. A large ad was placed in a Boston Newspaper in mid-May of 1838, offering for sale each of the stage routes with the equipment required to support those routes. Separately, the extra horses and vehicles with stable furniture, were also on the block. Real estate in Salem, Newburyport, and Dover was offered.

Lastly, shares in the Newburyport Turnpike, the Parker River Bridge, and the Ocean Bank were included.[45] A list of the Directors included:

Samuel Larkin

J. L. Thompson

S. Sweetser

Eben Moseley

Moses Newell

Henry Whipple

Thomas Haven

Jeremiah Coleman

In a too little, too late effort, New Hampshire took another look at extending the charter of the company for five more years on June 21, 1838.[46] Five days later it was rejected.[47]

In response, a newspaper article from June 1838 depicts the situation with regard to corporations and the New Hampshire government.

> **Corporations**
>
> Amongst the political humbugs of the day is the prejudice sought to be excited in the popular mind against corporations. For a long time this

43 New-Hampshire Gazette (Portsmouth, New Hampshire) August 8, 1837 p2.

44 Ibid, July 11, 1837 p2.

45 Columbian Centinel (Boston, Massachusetts) May 16, 1838 p4.

46 New Hampshire Sentinel (Keene, New Hampshire) June 21, 1838 p2.

47 New Hampshire Gazette (Portsmouth, New Hampshire) June 26, 1838.

> attempt was confined to the newspapers of the party, but latterly it has developed itself in our halls of legislation, where it threatens to interfere with the common and ordinary transactions of everyday life, and may become seriously mischievous unless the public mind be informed and so disabused in relation to the subject.
>
> If we refer especially to the unexpected and as we think unjust, refusal to grant the Eastern Stage Company further time, even so short as one year, to arrange its business, perform its contracts, and divide its property. No corporation within our knowledge has performed its duties towards the public in all respects so satisfactorily as this one; the manner in which it executed its important mail contracts was a topic of constant eulogy with the Post Office Department. It spared no pains or expense to afford facilities to the travelling public; and go where you would, the Boston and Portsmouth line of stages was universally spoken of as the best, as the pattern line, throughout the country. When therefore this corporation asked of the legislature an extension of its charter for the purpose of enabling it to execute its contracts for carrying the mails and for closing its business in the manor most for the advantage of all concerned, the House by a very handsome majority passed a bill for the purpose. But when the bill got into the Senate, by some unaccountable process, party politics got worked into action against it, and by a very strict party vote, the bill was thrown out, and every attempt to get the least extension, was sullenly resisted! The corporation is therefore extinct and must make the distribution of its property as it best is able. Why such an unusual severity was pursued towards it, that it should be thus made a sacrifice to the genius of Loco-Focoism, we are unable to surmise; but that it will be viewed with approbation by the people, we are slow to credit.[48]

Trustees took over the affairs of the company in June of 1838, and administered the company until 1840.[49]

Complaints about the new management emerged within months of the takeover.

> A writer for the Gazette complains of the stage companies for raising the fare to Boston. If such be the fact, it is deserving of censure; in hardly any other part of the country is the stage fare so high as it is here. The old Eastern Stage Company would never have been up to this trick of raising the fare as soon as the boat stopped; and it is one which will do them more harm than good after all, seeing that formerly they had the good will of the people, but if they persevere in that present high rates of fare they will certainly lose it.[50]

48 Portsmouth Journal of Literature and Politics (Portsmouth, New Hampshire) June 30, 1838 p2.

49 The Phillips Library at the Peabody Essex Museum, Eastern Stage Company (1814-1840) Records.

50 Portsmouth Journal of Literature and Politics (Portsmouth, New Hampshire) December 1, 1838 p3.

The trustees for the company tried to be innovative, thinking that if you cannot beat the railroad then maybe you should join them. They tried to match up their schedules to those of the railroads, by meeting a train which only went as far as Salem or Newburyport, then carrying the train passengers on to their final destination in Portsmouth or Portland. The stage line even offered to take anyone anywhere in the city they might want to go, and for only 12 1/2 cents.[51]

The Eastern Stage Company, or at least many of its employees, were remembered well into the 1840's Messrs. Ackerman, Annabel and Bancroft, who previously worked on the stages, became conductors on the Eastern Railroad.[52]

As late as 1848, John Mendum was remembered in the obituary column as a driver from the Eastern Stage Company who had made six thousand trips from Boston to Portsmouth, a distance equivalent to ten times around the world.[53]

While the Eastern Stage company was trying to get a foothold in the stage business and William Potter was offering "opposition lines," another new stage entity was forming in Maine. The Portland Stage Company was given a charter by the newly established Maine Legislature in March 1824.[54]

Within a few months, the company was offering an accommodation fare of $4.00 from Portland to Boston.[55] They also ran shorter routes such as Portland to Saco, Kennebunk to Dover, and South Berwick, Maine to Portsmouth. Conflict with the Eastern Stage Company seems to have been avoided and the two entities lived peacefully together in the market place. They were even named co-trustees of defendant Samuel Hodgdon, innkeeper of Portsmouth, in a suit for recovery of money due to Leonard Cotton of Portsmouth.[56]

Officers of the company over the years had been:

Nathaniel Mendum-Agent

Enoch Paine-Clerk

A. Rice-Director

Nathaniel Jeffords-Clerk/Director

Daniel Sewall-President

William Lord-President

Of course, their fate was also sealed by the advent of the railroads. By April of 1844, real estate belonging to the stage line was being offered for sale in Saco, Portland, Kennebunk, and Portsmouth.[57] Their charter had been surrendered in February of that year.[58]

Contemporary descriptions of the vehicles used for stage transportation are useful to appreciate the level of luxury involved in public transportation at that time.

51 Gloucester Telegraph (Gloucester, Massachusetts) October 19, 1839.

52 Newburyport Herald (Newburyport, Massachusetts) December 4, 1843.

53 Cape Ann Light and Gloucester Telegraph (Gloucester, Massachusetts) July 15, 1848 p4.

54 American Advocate (Hallowell, Maine) March 6, 1824 p2.

55 Gazette (Portland, Maine) June 8, 1824 p1.

56 New-Hampshire Gazette (Portsmouth, New Hampshire) August 6, 1839 p4.

57 Portland Advertiser (Portland, Maine) April 23, 1844 p3.

58 Maine Farmer (Augusta, Maine) February 22, 1844 p3.

Concerning the "Stage Wagon:"

> The body was rather long in proportion to its breadth and contained four seats, each holding three passengers who all sat facing the horses. From the height of the seats it was open all around. The roof was supported by slender shafts rising up at the corners and sides. In wet weather, a leather apron was let down at the sides and back, and fastened to buttons. The curtain in the front separated the driver from the passengers.
>
> The wagon had no door and the passengers got in through the front, stepping over the seats as they went to the end of the wagon. The driver sat on the front seat with a passenger on either side. The heavier luggage and trunks were fastened behind upon a frame, while smaller articles and the mail bag were stuffed under the seats, to the great annoyance of the passengers, who were frequently forced to sit with their knees up to their chins. One traveler remarks that the passengers' feet were insinuated between two trunks, 'where they are most lovingly compressed whenever the vehicle makes a lurch into a rut.'[59]
>
> The body of the wagon was suspended on two leather straps as were the early stagecoaches, passing lengthways under it and secured upon strongly propped horizontal bars in front and behind. There were no backs on the benches to comfort passengers during a rough and fatiguing journey over a newly and poorly constructed road. The first three in had the advantage of resting their shaken frames on the back of the wagon. Women were usually given these seats.[60]

In contrast;

> The American 'Mail Stage' in which we journeyed over so many wild as well as civilized regions, deserves a place at our hands. The springs, it will be observed, are of hide, like those of the French Diligence — and everything about it is made of the strongest materials. There is only one door, by which the nine passengers enter the vehicle, three for each seat, the center sufferers placing themselves on a movable bench, with a broad leather band to support their backs. Instead of panels, these Stages are fitted with leather curtains. The baggage is piled behind, or is thrust into the boot in front. They carry no outside passengers - and indeed it would try the nerves as well as the dexterity of the most harlequin that ever preserved his balance, not to be speedily pitched to the ground from the top of an American coach, on almost any road that I had the good fortune to travel over in that country.[61]

For one who has never had the opportunity to ride in a stagecoach, it would be impossible to

59 John M. Duncan, *Travels Through Part of the United States and Canada in 1818 and 1819.*

60 Thomas Twining, *Travels in America 100 Years Ago* (1795).

61 Basil Hall, *Forty Etchings from Sketches Made with a Camera Lucida in North America, in 1827 and 1828.*

conceive how passengers would feel, packed in like sardines. To that, we must add the bumping and swaying of the coach, even on firm and relatively smooth roads, the kind of roads that simply did not exist in that day. Of course, the gentlemen passengers might also be smoking cigars, while many would imbibe from a small bottle as the journey progressed. No, riding in a stage was not for the feint of heart!

Chapter Three

Taverns, Hotels and Public Houses Boston to Portsmouth

Taverns were plentiful on the road from Boston to Portland. Many were in place shortly after settlement began. At first, they were called "Ordinaries" and surprisingly, were even advocated by the Puritan forefathers. Of course, the licensing was tightly controlled, where only the most reliable men need apply. Apparently, even the most righteous of that time understood that after a day filled with sermons delivered in the coldest possible settings, a man might need to warm himself by the tavern fire with a glass of grog firmly in hand. For this reason, taverns were often found immediately adjacent to the church or meeting house.

Taverns of that day had multiple missions to serve. First, they provided a place to gather and exchange newsworthy stories. Most were also inns for serving the traveling public with libations and food, along with a place to sleep. The term "a place to sleep" did not mean a private room, and sometimes not even a private bed. Stories abound of travelers just getting to sleep when the inn keeper would wake them with news that another arrival would be coming to share the bed.

The best place to find which tavern was used by which stage line is to note the "Stage Notices" in the newspapers.

Usually, these advertisements would name the places where the stage books were kept, or at least mention the taverns where stages started their trips. We do know that stages stopped every nine to twelve miles to change horses and give the passengers a short respite from the hardship of stage travel. Many of these relay stations were not recorded in the newspaper advertisements unless they happened to be a point where a stage could be boarded by passengers.

As the road began in Boston, we will begin the tavern's stories there as well:

Lighthouse and Three Cranes Taverns—Boston

Very little has been recorded about the Lighthouse Tavern. All we do know is that it was located at State and Devonshire streets in the North End of Boston, and that Mr. Whatley was the innkeeper in the middle to late eighteenth century. Bartholomew Stavers used it as his headquarters after he got his stage line to Portsmouth up and running. Although he ran the stage business out of the Lighthouse Tavern, his stage line terminal was at Three Cranes Tavern in Charleston. His reasoning, possibly, was that since a stage to Boston would have to cross the Charles River by ferry, it would be easier to locate the livery stable and coaches on the Charleston side of the river and let his passengers worry about how to get from Boston proper to Charleston.

Three Cranes Tavern had an exceptional history. Originally it had been the Great House constructed for the use of John Winthrop in 1629.

> The Great House, was the first public building erected by the Massachusetts Bay Colony. The Great House was built as part of the Charlestown settlement in 1629 for Governor John Winthrop. As the Governor's residence and a meeting place for his Court of Assistants, the Great House was the Company's seat of government for several months during the summer of 1630. Governor Winthrop and the operations of the Massachusetts Bay company moved across the river to Boston (then called Shawmut) in October 1630. The company sold the Great House to the town in 1633.
>
> The Great House was then used as a meeting house and center of community life. It later became the Three Cranes Tavern when it was purchased by Robert Long in 1635. The Three Cranes Tavern served merchants, sailors, and craftsmen that required lodging, and provided a meeting place were ideas as well as goods could be exchanged.
>
> The buildings in City Square were razed by fire during the Battle of Bunker Hill on June 17, 1775. The square was left as an open marketplace after the battle and the remains of the Great House/Three Cranes Tavern site were preserved beneath the ashes.
>
> 'Time came to a standstill in Charlestown on June 17, 1775. While the battle raged on nearby Bunker Hill, the British Army set fire to the town itself. Flames swept through the heart of Charlestown and destroyed the Three Cranes Tavern. Much of the town was later rebuilt and then changed time and again, but the tavern site became City Square and has remained an open space in the city landscape ever since. This was no ordinary site, for the tavern had originally been Governor John Winthrop's 'Great House' in 1630 and briefly served as the seat of government in Massachusetts Bay Colony.'[1]

This tavern is first recorded in 1658 at the corner of State and Kilby streets as William Davis sold it to William Ingram. From there, it passed from John Holbrook in 1680, to Thomas Waite in 1731, to Simon Eliot in 1760, to Leonard Jarvis in 1769, to Joseph Rotch, Jr. in 1772, to Elisha Doane in 1773, and then to Isaiah Doane in 1786.

[1] Amy Laskowski, Boston University DIG, *Secrets of the Three Cranes Tavern*. Retrieved June 14, 2017, from BU.edu website: <http://www.bu.edu/today/2014/boston-archaeology/>

Bunch of Grapes Tavern—Boston[2]

In 1765, Colonel Ingersol, then innkeeper at the Bunch of Grapes, was hauled into a selectmen's meeting for having a billiard table at his establishment. Ingersol admitted that it was true, but hastened to add that two or three gentlemen had rented a room from him and had the table installed. He then stated that he did not realize that such an arrangement was against the law, but would now have it removed immediately.

The selectmen reminded him that if he did not comply, he would not be licensed for the upcoming year. No doubt the scene played out hundreds of times as the tavern keepers tested the authority of the local government. They could lose their license for the slightest impropriety that occurred at their establishment.

Josiah Paine's earliest stage line advertisement in 1800 records the Bunch of Grapes in Boston as the location of the stage books. At the time, Mr. Kendal was the innkeeper. The tavern was listed as being on State Street in Boston in 1805. Shortly thereafter it was demolished and Kendal opened a new tavern in Medford.

A sign on a building near the original location for the tavern states:

> **BUNCH OF GRAPES TAVERN**
> Here on 30th of July 1733 was instituted under charter from the Grand Lodge of England the first regularly constituted lodge of Free and

[2] From the collection of Christopher Hodapp.

Accepted Masons in America now St. Johns Lodge of Boston.

Here in 1786 was organized the Ohio Company, pioneer in the development of the great west under the leadership of General Rufus Putnam. First township laid out at Marietta Ohio.

The first recorded owner of this tavern, located at the corner of State Street (then King Street) and Kilby Street (then Mackrell Lane), seems to be William Davis. He sold to William Ingram in 1658, who conveyed this property, as 'The Bunch of Grapes ' to John Holbrook in 1680. Then a wooden building, it came into the possession of Francis Holmes about the year 1709. It burned down in 1711, when much of Boston was destroyed. It was replaced in 1712 by the brick structure shown in the painting reproduced above. One of the earliest, as well as one of the most famous, of Boston taverns throughout the land, it was called by many 'the best punch house in Boston.'

Two o'clock was the hour at which all in the town of Boston dined in the early days, and guests were called together for the meal by the ringing of a bell in the street. The customary fare at the Bunch of Grapes was salmon in season, veal, beef, mutton, fowl, ham, vegetable, and pudding, with a pint of Madeira before each partaker. The carving was done at the table, in the old English way, each guest helping himself to what he liked best. Five shillings a day was the usual charge.

Successors to Holmes as landlord were; William Coffin, 1731; Edward Lutwych, 1733; Joshua Barker, 1749; William Weatherhead, 1750; Rebecca Coffin, 1760; and in 1790 James Bowdoin is recorded as bequeathing to his wife a house called 'The Bunch of Grapes.'

As the Revolution grew near and lines came to be drawn between Bostonians, the Bunch of Grapes became the resort of the High Whigs, who made it their political headquarters. With military occupation, the tavern was kept by Salem-born Captain John Marston during the years of 1775-1778. A cold welcome was met by scarlet uniforms there, and this atmosphere led the latter gentry to meet at other taverns where the talk and toasts were more to their liking.

Prior to the Revolutionary period, on Monday, July 30, 1733, the first Provincial Grand Lodge of Masons in America known as Saint John's Grand Lodge, was organized at the Bunch of Grapes by Henry Price, a Boston tailor, who had received authority to do so from Lord Montague, Grand Master of England. Price was also authorized to 'constitute one or more regular Lodge, or Lodges, as he shall see fit.' At the same meeting during which the Grand Lodge was formed 'at the House of Edward Lutwych at the sign of the Bunch of Grapes in King Street, Boston, New England,' Price was petitioned by those present to constitute them into a regular Lodge. The petition was granted and the First

Lodge, often called the Mother Lodge, was formed, with Henry Hope as the first Master. For the first half century, the records of the Grand Lodge and the First Lodge are intermingled, and the transactions of one body are also the records of the other. In 1783, the First and Second Lodges, the latter founded in 1761, successfully petitioned the Grand Lodge to be united under the name, title and designation of Saint John's Lodge, A.F. and A. M. Price became Grand Master for the second time on October 23, 1767. A monument now marks his grave in the cemetery in Townsend, Massachusetts, bearing the following inscription:

HENRY PRICE
Founder of
DULY CONSTITUTED
MASONRY IN AMERICA

Meetings were held by these pioneer Masonic groups at other Boston taverns, including the Royal Exchange, on the site now occupied by the Merchants National Bank building on State Street (then King Street). Luke Vardy, a brother of the order, was the landlord and, according to a bit of verse credited to Joseph Green, which was first read at a meeting in 1749,

'Twas he who oft dispelled their sadness, And filled the Brethrens' hearts with gladness. For them his ample bowls o'erflowed, His table groan'd beneath its load; For them he'd stretch'd his utmost art Their honors grateful they impart.'

Among other taverns where they met were the Concert Hall on Hanover Street and the Greyhound in Roxbury.

Upon the evacuation of Boston by the royal troops, the Bunch of Grapes became the center of popular demonstrations. General Washington was handsomely entertained there. On July 18, 1776, after hearing for the first time the Declaration of Independence read from the balcony of the Town House at the head of State Street, the populace proceeded to pull down from all public buildings the royal arms which had distinguished them. They made a bonfire of these trophies in front of the Bunch of Grapes.

To celebrate General John Stark's victory at Bennington, a banquet was held at this famous old inn. Two brass field-pieces were set up in the street and fifers and drummers were stationed on the balcony of the Town House. The celebration opened with a salute of thirteen rounds discharged by the cannon, and at every toast, three rounds were fired and a flight of rockets was sent up. Around five o'clock, two barrels of grog were brought out into the street for those who had collected there. In spite of all the toasts and free grog, a contemporary writer stated, 'It was all conducted with the greatest propriety, and by ten o'clock every man was at his home.' Shortly after this General Stark arrived in town and was entertained royally at this popular tavern. It was at that time that he presented the trophies of war which now adorn the Senate Reception

Room in the State House on Beacon Hill.

On his return from France in 1780 Lafayette was received at the Bunch of Grapes with all honors, as he brought the news that France had at last cast her powerful weight into our Revolutionary contest.

Perhaps one of the most important events to take place here, as far as our nation is concerned, was the 1786 organization of the 'Ohio Company of Associates,' generally known as 'The Ohio Company,' by a group of veteran Continental officers called together by Dr. Manasseh Cutler and General Rufus Putnam. The settlement of the great State of Ohio began in earnest in December of the following year under the leader-ship of the Company superintendent, General Rufus Putnam. Putnam was a cousin of the better known hero of Revolutionary fame, General Israel Putnam of Salem Village (now Danvers), Massachusetts. It was the latter, then a Major, who like Cincinnatus of old, left his plough in the furrow and hastened to Cambridge on hearing the news of the fighting at Lexington and Concord in April of 1775.

The famed Trinity Church of Boston was also organized at the Bunch of Grapes. It is apparent that this renowned hostelry of our earliest days merits a high place in the annals of Boston.[3]

[3] Allan Forbes and Ralph M. Eastman, *Taverns and Stagecoaches of New England.*

Eastern Stage House—Boston

This was the home office in Boston for the Eastern Stage Company at 84 Ann Street. Its name seems to first appear when the Eastern Stage Company began in 1818 and lasted until the building was demolished in favor of some granite shops in July, 1846. During that period, it served its owners well. One of its striking amenities was the inclusion of two large stables capable of housing 100 horses.

In 1833, a letter to the editor of the Newburyport Herald stated that the paper had failed to mention the Eastern Stage House in an article concerning the "striking improvements in the public houses of Boston." He reminds the editor that the Old Eastern Stage House had been newly fixed up and furnished with numerous conveniences added. Having been so prominent in the affairs of a city the size of Boston, it is no wonder that the Eastern Stage House became the setting for some bizarre happenings.

One such event was the supposed robbery and murder of Mr. Johnathan Perkins who disappeared shortly after leaving his son and a hired man at the Eastern Stage House. Perkins had traveled from Newington, New Hampshire, where he was postmaster to Brighton, Massachusetts with 20 head of cattle and 500 sheep. After selling them, he proceeded to check in at the Eastern Stage House. Soon thereafter, he told his son that he had some business to conduct in Cambridge and might not be back until after ten. He also stated that he would walk back if he could not find any suitable transportation. Around ten o'clock that evening, on the Old Cambridge Bridge, a passerby found Perkin's hat and wallet which was without cash and contained only identification papers. A wide search began, but of all his searchers, his creditors knew best where to look. A few days later, he was found working on a farm in Rhode Island and needed little persuasion to repay his creditors and have all the charges against him dropped.

Newspapers up and down the East Coast carried the story, but the American and Commercial Daily of Baltimore had the best summary.

> **Murder, no murder.** When the man at the east wishes to disappear, it has come to be the fashion for him to leave his lodging's in Boston, saying that he's going to Charleston, and may not be back until late in the evening. The next morning his hat and pocket-book (with no money in it) are found on one of the bridges and the story is told in the newspapers under the head of supposed murder. The man's creditors, however (and he is sure to leave such relations to mourn for him) immediately set themselves to inquire which way he went after the murder scene. The last tragedy of this sort was played out by Mr. Jonathan Perkins, postmaster, Newington New Hampshire, who, a few days afterwards was found at work on a farm in Newport, Rhode Island none the worse for having been murdered![4]

Less than a year later, his wife Martha had filed for divorce based on her husband having run away with one Ruth Adams, to parts wholly unknown.[5]

In 1838, just before the demise of the Eastern Stage Company which owned the house, Mr. J. Tyler acquired the Eastern Stage House. In November of 1838, Tyler advertised that the house had undergone complete repairs and had been refurnished throughout. It was then reopened for business, still keeping the books for the Eastern Stage Company but adding to that the books for the Eastern Railroad.[6] Perhaps in desperation, Tyler advertised new low prices for boarding at the house. Full meals were $1 per day for a single day, 85 cents per day for two days, or $5.50 per week. He attributed this to the current lower price for provisions but one wonders if he was really so generous as to pass those lower costs on to his customers.[7]

By 1841, John S. Robinson had taken over the Eastern Stage House.[8] Over the next year on at least two occasions, customers had left without paying their bills and in possession of clothing belonging to other tenants. Apparently, the level of clientele had significantly lowered in the years after the Eastern Stage Company had left.[9]

4 American and Commercial Daily Advertiser (Baltimore, Maryland) October 29, 1834 p2.

5 New Hampshire Patriot and Sate Gazette (Concord, New Hampshire) July 6, 1835 p4.

6 Newburyport Herald (Newburyport, Massachusetts) March 5, 1839 p4.

7 Exeter News-Letter and Rockingham Advertiser (Exeter, New Hampshire) November 10, 1840 p4.

8 Portsmouth Journal of Literature and Politics (Portsmouth, New Hampshire) September 11, 1841.

9 Newburyport Herald (Newburyport, Massachusetts) October 12, 1841 p2.

Lynnfield Hotel-Lynnfield[10]

July of 1846 found the house at the point of being demolished. The roof had already been torn off. Its demise was credited to the building of the railroad which had left the house essentially deserted. One old stage driver, Jack Mendum, had lived there for thirty or more years.

Lynnfield Hotel—Lynnfield

In 1808, shortly after the completion of the Newburyport Turnpike, an announcement appeared in the Columbian Centinel newspaper, announcing the opening of the Lynnfield Hotel. "Ebenezer Lewis respectfully informs his friends and the public, that he has taken the hotel in Lynnfield, on the Newburyport Turnpike, about 10 miles from Boston, and every exertion will be made to keep a good house of entertainment, where travelers and parties of pleasure will be waited on at all hours, and every favor gratefully acknowledged. There is a fine pond for fishing and a commodious boat provided. Sportsmen will always find plenty of game in the vicinity of this place. Good stables for horses, and a faithful and attentive groom is engaged. A few gentlemen borders can be accommodated and with chambers to themselves."[11]

Thomas B. Wellman, in his book *History of The Town of Lynnfield, Massachusetts*, relates some points of interest concerning the hotel.

> The old hostelry was built in 1804, by the old Newburyport Turnpike Company, who laid out the Newburyport Turnpike, which was finished at an expense of $420,000 in 1806. The hotel, 90 years old, was much better known than the town

[10] Courtesy Lynnfield Historical Society.

[11] Columbian Centinel (Boston, Massachusetts April 30, 1808 p4.

itself, and now nothing remains of the famous building but ruins.

It was sold in 1832 to a man by the name of Thomas Pingree of Salem, having been run by landlord Chester for two years. Other proprietors were Noah Newhall, Asa and Theron Palmer, Capitan Smith, Charles Putnam, Mr. John Andrews, Messrs. Fisher and Sprague, and Samuel S. Bruce, among others

Dr. Perley bought it and used it as a private residence until his death. It covered an acre and contained about 60 rooms.

The street by Lynnfield Hotel was double width for one-half mile, and it was said that it was built by the company for the farmers to trot horses upon. So well-known is the building that the village was called 'Lynnfield Hotel' more commonly than anything else. With its French roof and number of stories, it was the most imposing building perhaps in town.

The Lynnfield Hotel has caught fire many times, but never proved serious until the night of March 7, 1894, when the fire caught in the kitchen. Within an hour from the time it took fire everything was flat though everything possible was done to save it. Two other hotels were built at the same time as this, one at Topsfield and one at Newburyport. We remember to have heard it said years ago that the cost of the Lynnfield Hotel was $22,000 when built. No one was injured at the fire.

At the time the hotel was burned, an article appeared in the Salem Paper from which we copied the following facts. In 1855 the place was leased to Messrs. Hill and Townsend. The house was kept mostly for summer borders, and was generally well filled. Among the families boarding there at this time were those of Dr. Henry Millis, Mr. Theodor Neal, and Henry Poor of Salem. There were some wealthy Lynnites also who were also frequent visitors. They spent their money lavishly, and no doubt enjoyed themselves accordingly.

In November of 1856, Jack Saunders took the place. It was while the hotel was under his management that the military muster was held in Lynnfield. It is still remembered by the old residents as a reign of terror for the three days of its continuance. The hotel was converted into a veritable gambling hell, and every place in the vicinity which could be secured was turned into a den of infamy. The place swarmed with the dissolute of both sexes. Nothing occurring during the whole encampment, from 1861 to 1862, that can compare with these three fearful days.

In the earliest days of the hotel a murder was said to have been committed there, and subsequently a ghost story was connected with the same. An old lady, long since dead, who had lived in the vicinity all her life, told the writer all the particulars of the affair, avowing her firm belief in the ghost part of it. She knew of persons who lived at the hotel who had seen the ghost with their own eyes and almost been frightened out of their senses. This old lady also stated that she knew the person who wore out the clothes that were supposed to have been taken from the body of the murdered man, and gave his name. Although the writer was inclined to feel a little skeptical about some parts of this narrative, he could not be positive that it was not true.

The fire at the hotel was seen for many miles around and was at once a source of joy and sorrow for the immediate and surrounding community. The ruins still speak of what once was, and now is not.

One survivor of the hotel is still in existence. In one of the halls of the Essex institute of Salem is the ancient tavern sign of 83 years ago. It has upon it a life-size picture of General Washington handsomely painted. Above it says 'Washington Hotel' and below 'Pete Newcomb, 1812'.[12]

Topsfield Hotel—Topsfield

George Francis Dow in his *History of Topsfield, Massachusetts* sets out some interesting points about the Topsfield Hotel.

When the turnpike was built from Boston to Newburyport in 1803 to 1804, the Newburyport Turnpike Corporation erected a hotel on the rise at the junction of what is now High Street, near the site of the house built by Daniel Perkins and still owned by his descendants. It was expected there would be heavy travel all over this highway, and the Topsfield hotel soon became the stage center of Essex

Only Known Sketch of the Topsfield Hotel[13]

County, and was said to have been the best tavern on the Eastern Stage route. It is not known who was first in charge, but Ephraim Wildes probably leased it when he was given an inn holders license in 1807. In 1809, he advertised in the Salem Gazette that he had taken the hotel and solicited the patronage of the public. The next proprietor was Cyrus Cummings, who kept the hotel for a number of years. The records show he was first given a license in 1812.

In 1808, the hotel held the famous Caucus to denounce the embargo, and in March of 1812, the court martial for the trial of Capitan Samuel Griffen of the 5th regiment, 2nd division. Here, in 1818, the Essex Agricultural Society was organized, and its annual meetings where held here for a number of years. In 1829, the Confederation of County Lyceums was organized and in 1834, the Essex County Natural History society, from which grew the Essex Institute. Its parlors were the scene of many political and social gatherings.

[12] Thomas B. Wellman, *History of the Town of Lynnfield, Massachusetts.*

[13] From a pencil sketch by Alonzo Lewis of Lynn, *Historical Collections of the Topsfield Historical Society* Vol VIII.

The hotel was not profitable and in 1823 it was sold at auction for $3,135 to Ephraim Wildes who with his Brother Solomon and nephew Moses later ran a widely known hotel for stages in Boston. On December 15, 1826, Cyrus Cummings, who was still the proprietor, bought the property from Mr. Wildes for the same sum. In an advertisement in the Salem Gazette on July 13 1824, Mr. Cummings stated that this hotel 'offers a most agreeable and healthy summer residence to invalids and others who are disposed to spend the summer months in the country. The bar is always supplied with the choicest liquors, and the table with the best provisions of the season. Parties of pleasure and others may depend on respectful attendance and every effort will be used to give perfect satisfaction to all who favor the establishment with their patronage.' Mr. Cummings continued as proprietor of the hotel until his death in 1827, at the age of 45 years. His wife, Susan, continued the establishment until 1835. That year she advertised the property for sale setting forth its many advantages and announcing that 'seven regular stagecoaches stop at the hotel every weekday and that private travel has been constantly increasing.' The purchaser was John Holland, who for many years was known to travelers on the Eastern Stage as one of the most intelligent, obliging and temperate of that body of drivers.

The stage lines were soon discontinued, for the coming of the railroad had taken the through travel and although the railroad did not extend to Topsfield until 1854, the patronage at the hotel was so poor that Mr. Holland was forced to mortgage the property to Solomon and Moses Wildes, who took possession of it in 1840. They advertising for sale or to let on June 1 1841, it then being vacant. It might be interesting to include the description of the place. They said it was 'one of the pleasantest situations for a summer residence that can be found, being very airy. The house is three stories high, has a fine hall and was built expressly for a public house. Adjoining is a large stable, sufficient to accommodate 50 or 60 horses overnight, together with large wood sheds and other outbuildings. There are about three acres of first rate land, on which is about 100 grafted fruit trees of select fruit. One of the finest wells of water, together with an aqueduct that never fails of water.'

Joseph Bassett, shoe manufacturer, and William Fabens, lawyer, both of Marblehead, bought the building, and in 1844 it was removed to Clifton. The venture was not a financial success and on January 1, 1846, it was totally destroyed by fire.[14]

Judging from the images of these two hotels, they were probably designed by the same architect. Both were three stories, had similar roofs with a cupola at the top. This is not surprising as the Turnpike Association built both when they constructed the turnpike.

[14] George Francis Dow, *History of Topsfield Massachusetts.*

The First General Wolf Tavern— Newburyport[15]

In 1762, Captain William Davenport opened his doors under the sign of General Wolfe. Sturdy patriarchs sat around the fire in Davenport's parlor, nursing their toddies while rocking the cradle of liberty.

The Sons of Liberty rallied volunteers for the Continental army. They raised their glasses to toast Nathaniel Tracy and his 'Yankee Hero' the first privateer outfitted to fight British fleets. During the Revolution General Lafayette used The Wolfe for his headquarters.

The original Wolfe burned in the Great Fire of 1811, and a second opened on the corner of Harris and State streets.

Dueling rose up in the early 19th century and Bostonians needed a refuge from the authorities. Seabrook became an arena to settle scores. The battle was usually over an object, namely a woman.

The duelers each had a doctor and umpire on hand. As a rule, the crew lodged at the Wolfe on the eve of battle. They crowded in the lounge belting down liquid courage. At early dawn, they would work their way over the border to some secluded area to 'conduct their bloody business.'

In 1809, a duel set to go down between a Harvard grad Army General and a South Boston fellow made headlines. It was ignited when a Boston maid would give neither the satisfaction of who she fancied. However, the day before the match she sent a red rose to 'Southie.'

[15] John J. Currier, *The History of Newburyport.*

'It smells of disdain!' cried the General. He never showed to fight. He was spotted at the Wolfe on 'the hot seat' drinking and thinking his maid away.

In 1828, the Eastern Stagecoach Company purchased the Wolfe Tavern. The company flourished for 20 years until railroads reduced the need. However, a few of the 'Whips' refused to hand over the reins.

'Old Major Samuel Shaw put on a coach with the fare at $1.50 to Boston. David Batchelder ran a coach from Amesbury to the front door of the Wolfe. John G. Whittier put great stock in coach service. Guests at the Wolfe would receive the following note: 'I'll get my parcel or you will get your books if left in care of Mr. Batchelder.'[16]

When Newburyport was incorporated in 1764, Wolf Tavern, then standing on the corner of Fish (now State) St, and Threadneedle Alley, was a famous resort for travelers. The proprietor, William Davenport, came to Newbury as early as 1738, and established himself in business as a woodcarver. He married Sarah, daughter of Moses and Mary (Noyes) Gerrish of Newbury on April 3, 1740. In 1741, he purchased land from Samuel Greenleaf on Fish (now State) St., with a dwelling house thereon, which he owned and occupied until his death in 1773. His father, James Davenport, was a prominent innkeeper in Boston.

William Davenport had command of a company in the expedition that sailed from Louisbourg in June 1759, for the reduction of Quebec City. He was in the Battle on the Plains of Abraham on the 13th day of September, when General Wolfe was killed. After his return to Newbury, he converted his dwelling house into a tavern and hung out a quaint sign with the head and bust of General Wolfe carved thereon.

Sign for the Wolfe Tavern[17]

In 1764, a part of the town of Newbury was set off and incorporated by the name of Newburyport. Wolf Tavern was then near the business center of the new town. Social and political gatherings were held in its parlors and the Newburyport Marine Society, as

[16] Newburyport News (Newburyport, Massachusetts) February 28, 2015. Used by permission of Melissa D. Berry, Author.

[17] John J. Currier, *The History of Newburyport, Massachusetts 1764-1905.*

well as the members of Saint Peter's lodge of Free and Accepted Masons, met there as often as once a month for many years. The mail stage stopped at the tavern door long enough to allow the traveler to quench his thirst at the bar, or appease his hunger with a substantial meal. Farmers from New Hampshire, bringing eggs, butter and poultry to market, frequently found shelter for the night under its roof. William Davenport died September 2, 1773. Under the management of his sons, Anthony and Moses Davenport, the tavern continued to be a popular and prosperous place for more than 30 years. In 1782, Marquis de Chastellux, who served under Count de Rochambeau in the Revolutionary war, visited Newburyport and subsequently published an account of his journey through Massachusetts and New Hampshire. He dined at Wolf Tavern, or Davenport's Inn, as it was frequently called and remained there overnight with his traveling companions. At a later date, Governor Hancock and other men prominent in public life were entertained at this famous hostelry. In 1804, Thomas Perkins took over the helm and renamed the tavern the Perkins Inn. By 1807, Prince Stetson took over what was once again the Wolfe Tavern. Four years later, in May of 1811, fire destroyed the inn and much of Newburyport. Stetson tried to reopen by leasing the home of Colonel Stephen Bartlett but by 1814 had abandoned the project.[18]

Merrimack Hotel in Newburyport

In December of 1813, Benjamin Hale, one of the owners of the Portsmouth, Newburyport and Boston Stage company, purchased a large three-story brick house, built by John Peabody in 1807, on the corner of Harris and State street, and fitted it up for the accommodation of travelers. It was called the Merrimack Hotel. 'Lately the mansion house of General Peabody, at the Sign of James Wolfe, Esq, on State Street; Stephen Hammond, landlord.' The portrait of General Wolfe on a swinging sign, painted by Moses Cole, an artist of some distinction at that time, was placed in front of the house, where it still hangs.[19]

Benjamin Hale did things a little differently than other stage line owners. Instead of starting a stage line to bring more customers to his tavern, he opened a tavern for the convenience of his stage line customers.

Wells' Stage House / Langmaid's Tavern House-Hampton Falls

Hampton Falls was the stage town where the horses were kept and the changes made. People came from other places to take the stage here, a great many coming from Exeter. From 100 to 125 horses were kept here at one time. Caleb Tilton had charge of the stables. The keeping of so many horses here was a great advantage to our farmers, as it made a good

[18] John J. Currier, *History of Newburyport, Massachusetts 1764-1905.*

[19] Ibid.

demand and a ready cash market for hay and grain at good prices.

At first the stage house was kept by Capitan Joseph Wells, and later by his son Moses, in the house now occupied by Miss Abby Gove. After the tavern house was built in 1808, that was the stage house. The house was kept for 15 years after it was opened by Edward Langmaid, and afterward by a number of others until the stages ceased to run in 1840.

Under the management of Captain Wells and his wife Eunice, the house became celebrated for its excellence. It was called the best public house between Portland and Boston. The traveling public made their plans to remain here overnight as much as possible. From all we can learn this was the best public house ever kept in the town. Not a little of the credit was due to the good management of Mrs. Wells. Captain Wells built the house now occupied by Miss Gove in 1787, and kept a tavern there until his death in 1791. The house was kept after his death for a few years by Mrs. Wells, who was succeeded by her son Moses. Mrs. Eunice Wells removed to Newburyport, and was taxed here as a nonresident until her death in 1831. Moses Wells continued to keep tavern here until the tavern house, so called, was built in 1808. This ended the career of what had been known as the Wells' Tavern. Moses Wells died in 1825.

The new house was kept by Edward Langmaid, who came from Chichester and had previously assisted Mr. Wells

Hotel Whittier-Hampton[20]

[20] Courtesy Hampton Historical Society.

> in the management of his house. Mr. Langmaid continued to keep the house until 1823. He was the first postmaster of the town. During his occupancy of the house, it became the stage house and continued to be such as long as the stages ran. After leaving here, Mr. Langmaid returned to his farm in Chichester. After he went away, the house had a number of different landlords,—Powell, Leavitt, Mathew Mariam, and lastly, Newman Brown. After the stages ceased to run this house was not used as a public house, but was rented, having one or more families living in it.[21]

Leavitt's Tavern / Whittier Hotel-Hampton

Beginning before 1750, a public house opened by Jonathan Leavitt was in place in Hampton, New Hampshire (on today's Route 1). Joseph Dow in his book, *History of the Town of Hampton, New Hampshire*, tells the story of the tavern:

> Lieutenant Jonathan Leavitt kept tavern many years, where the hotel in the village now stands. The date of opening is not preserved; but he was married in 1735, and very possibly began, not far from that time. A tavern was continued without interruption, by his son Thomas Leavitt. He died in 1791, leaving a widow, who still carried on, till their daughter, Nancy, was married to Josiah Dearborn, about 1803; after which, Mr. Dearborn was for many years, proprietor and landlord. He built the present house, probably about 1816 or '17.
>
> The new house was called Josiah Dearborn's Inn, and in later years Mr. Dearborn, having left his elder son in possession, had gone to blacksmithing on the adjoining lot. It was then designated as Samuel Dearborn's Inn. A short time previous to 1840, Oliver Godfrey and the late Alfred J. Batchelder leased the house, and named it The Railroad House, in honor of the railroad, then in course of construction.
>
> By this name it was known for some years, under different landlords, all leasing of the Dearborns.
>
> In 1859 or 1860, Jared P. Whitcomb bought the premises, repaired, repainted, and gave it the name Union House, by which the hotel has ever since been called, until it was changed to the present name in 1890.
>
> Mr. Wittcomb was succeeded in the proprietorship by Bean and Whittier, in 1864, — this firm, by Smith and Whittier; —and for the past ten or twelve years, Mr. Otis H. Whittier has been sole proprietor. He built a large wing in 1886, and the hotel can now accommodate from 80 to 100 guests. In 1917 the house burned down. The cause of the fire according to the town report for that year was 'rats and matches.'[22]

While the principal stop for the stage at Hampton was at Jonathan Leavitt's until 1802, after that time, a relative's,

[21] Warren Brown, *History of the Town of Hampton Falls, New Hampshire.*

[22] Joseph Dow, *History of the Town of Hampton, New Hampshire.*

Earl of Halifax Tavern-Portsmouth[23]

James Leavitt's, tavern was the preferred stopping place. James had bought the mansion house of General Jonathan Moulton, well known as the 'Faust ' of New England. Moulton had made an agreement with the devil but had tried to trick the devil so that the devil burned down Moulton's first house. Moulton cleverly rebuilt on the same property, but moved the foundation just enough to avoid the old cellar. It was this second house which Leavitt used as his tavern up until about 1840.[24]

John Stavers, with his brother Bartholomew, built the Earl of Halifax Tavern in the early 1760's and operated it in conjunction with their line of stages from Portsmouth to Boston. Business being good, he planned a larger building which was constructed in 1766.

Henry Wadsworth Longfellow spoke of the original Earl of Halifax Tavern in his poem "Lady Wentworth."

[23] Free Masonry-Historic Houses.

[24] Cheryl A. Lassiter, *A Meet and Suitable Person*.

One hundred years ago,
and something more,

In Queen Street, Portsmouth,
at her tavern door,

Neat as a pin, and blooming as a rose,

Stood Mistress Stavers in her furbelows,

Just as her cuckoo-clock
was striking nine.

Above her head,
resplendent on the sign,

The portrait of the Earl of Halifax,

In scarlet coat and periwig of flax,

Surveyed at leisure all
her varied charms,

Her cap, her bodice,
her white folded arms,

And half resolved,
though he was past his prime,

And rather damaged by
the lapse of time,

To fall down at her feet and to declare

The passion that had
driven him to despair.

For from his lofty station he had seen

Stavers, her husband,
dressed in bottle-green,

Drive his new Flying Stage-coach,
four in hand,

Down the long lane,
and out into the land,

And knew that he was far upon the way

To Ipswich and to Boston on the Bay![25]

[25] Henry Wadsworth Longfellow, *The Poetical Works of Henry Wadsworth Longfellow.*

The new building included a third floor, which was used by the Masonic Lodge for their meetings. Of course, when they were not meeting, Stavers could rent those rooms to the traveling public. Just before the start of the Revolutionary War, John Stavers renamed his tavern the William Pitt. Pitt had championed the cause of the colonists and the renaming effectively moved John Stavers from the Tory camp to the side of the colonists. Brother Bartholomew took a different course returning to England where he spent the rest of his life.

A fascinating story of John Stavers and his experience with the colonists one night in 1775, is told in a small publication bearing the title *The Portsmouth Book*. On the high post, near the north-east corner of the tavern, was put up the favorite sign of the Earl of Halifax. The sign was torn down by a mob of patriots in 1775, who regarded the hotel as under British influence. Stavers escaped and fled to Stratham, but on his return to Portsmouth two weeks later, he was seized by the Committee of Safety and confined in jail at Exeter. Afterward he took the oath of allegiance, and soon had all suspicions of Toryism removed.[26]

Instead of being a meeting house for the Tory faction, the Pitt Tavern became a meeting place for patriots. This tavern also had its long list of famous visitors, including John Hancock, General Henry Knox, the Marquis de Lafayette, the future King of France, Louis Phillipe, and George Washington. Today, the tavern is a part of the Strawberry Banke Museum and has been restored to its original state.

[26] C. A. Hazlett, *The Portsmouth Book.*

Bell Tavern—Portsmouth

Again from *The Portsmouth Book*: "The Bell Tavern on the north side of Congress Street, next to the West of Congress Block, was built in 1743; and in front of a post hung a sign of a bell. Here was the headquarters of the patriots. A guest wrote: 'It was not a beautiful structure. An architect would not hold it up as a model. It had no stately columns or pillars, dome, or tower; but it had a history and hallowed memories, which are more significant and enduring.' In 1852, the building was sold and made into three stories; and the old sign-post, with its blue bell, was cut down. The fire of 1867, swept the building away."

Mary Caroline Crawford in her book *Among Old New England Inns*, tells us about the early days of the Bell Tavern in Portsmouth, New Hampshire:

> The transition of the Packer house into 'Colonel Brewster's Ta'an', as Washington called the place in his diary, is interesting. Mrs. Packer had a mania for building extensions to her residence, and whenever her husband was absent on his duties of office she improved the opportunity to add a room or two. The house was thus enlarged until it had the proportions of an inn, and Colonel Brewster was glad to take it over as such in 1786.
>
> This was by no means the colonel's first essay as a landlord, however, for he had long successfully exercised that function at the Bell Tavern, a house which had been built in 1743 by Paul March and which displayed from a post in front an attractive bell, painted blue. Here the patriots were want to congregate while the Tories were making merry at the Earl of Halifax Tavern, drinking as long, if not so deeply, as their red coated rivals. During the period of Colonel Brewster's incumbency of the Bell Tavern, the Marquis de Chastellux was here entertained and in the published account of his travels he speaks of Mr. Brewster as 'a very respectable man, and much attached to his country.'
>
> After Colonel Brewster had transferred his interest to the Packer House, Jacob Tilton became the host at the Bell.[27]

By 1802, Seth Walker had replaced Mr. Greenleaf as the host at the Bell Tavern.[28] After a long and varied career, the Bell Tavern was swept away by fire in 1867. One of the landlords at the Bell had been a Mr. Purcel, one whose widow afterwards kept the Portsmouth boarding-house with which the career of John Paul Jones is bound up.[29]

Shannon / Hight Tavern-Portsmouth

Elizabeth P. Nowers, in her book *Portsmouth Women: Madams and Matriarchs Who Shaped New Hampshire's Port City*, provides us with an in-depth insight into a wife's involvement in running a tavern. In 1746, Alice Frost Shannon Hight and her husband Nathaniel purchased a lot on the Piscataqua River, near the ferry landing that connected Portsmouth to Kittery,

[27] Mary Caroline Crawford, *Among Old New England Inns.*

[28] Republican Ledger (Portsmouth, New Hampshire) October 19, 1802.

[29] Ibid.

Stoodley's Tavern—Portsmouth[30]

Maine. Slavery was still in vogue and a negro man, Prince, and a woman named Diana were on site to help. Alice surely oversaw the kitchen and housekeeping and probably served beverages in Nathaniel's absence. She may even have done some book keeping.

In 1753, Nathaniel died at age thirty-eight leaving Alice to support herself and four children. An unusual circumstance, both wife and husband's name being on the deed for the property allowed Alice to take over the tavern. Knowing the trade well, she applied for a license to run the place. The selectmen approved her application and Alice, as many other widows had done before, had a thriving business to watch over, along with her four children.

In 1754, Alice remarried to Charles Hight of Portsmouth a widower and sail maker with five children of his own. Although she turned the legal control of the tavern over to her new husband, Alice now had some tavern responsibilities, in addition to a family of nine children, plus one from her new marriage. Even with this huge responsibility, the Hight family prospered. Charles was even being paid to operate the ferry to Kittery.

Charles died in 1761 from "pleurisy fever." Now Alice had ten children, hers and her late

[30] Library of Congress Reproduction LC-D4-19869.

husband's, a tavern, a ferry boat operation, and other real estate to manage. At least some of her children were now grown up and could contribute to the family support.

Alice continued to run the tavern until her death in 1780, a sterling example of a woman who could not only provide for herself, but also provide for the needs of a large family. Nowers closes her story with the comment: "The fact that she was successful was a combination of her good fortune and ability!"[31]

> On the north side of Daniel Street, east of Penhallow Street, stands a gambrel-roof house, projecting into the street a dozen feet in front of the adjacent buildings. It was built and occupied by Colonel James Stoodley, and for many years the tavern was the stopping - place for travelers journeying from Massachusetts to Maine. The New Hampshire Gazette of that period contained the quaint advertisements of the stage to Falmouth, now Portland, which started from Stoodley's.
>
> In the third story was an arched hall used for Masonic purposes and for public dance parties. It was here that the amateur violinist, Colonel Michael Wentworth, second husband of Martha Hilton, often 'fiddled till morning on his favorite violin.'[32]
>
> Colonel James Stoodley fought as one of Rogers Rangers during the French and Indian War. At the same time, he owned his first tavern, the King's Arms, starting around 1753. When a fire destroyed his original tavern on January 25, 1761, he immediately built this establishment known as Stoodley's Tavern. James and Elizabeth Stoodley lived here with their children, Elizabeth and William, as well as two slaves named Frank and Flora. The tavern and inn became a popular Portsmouth gathering place. The upper story with the dormer windows is a spacious arched hall that also served as, a ballroom, and function room. Slave owners held auctions here in 1762 and 1767. Before and during the Revolutionary War, American patriots congregated at Stoodley's Tavern. It became the most popular stopping place for stagecoach passengers traveling between Boston and Maine.
>
> After Colonel Stoodley died in 1779, his widow remarried and boarded lodgers in the house. Colonel Stoodley's daughter, Elizabeth, married Elijah Hall, and they inherited Stoodley's Tavern around 1786. Hall was a shipwright who served as one of John Paul Jones' lieutenants on the Ranger and was aboard when the Ranger captured the Drake in 1778. Later, he was the State Councilor and Naval Officer for the Portsmouth district. When Elijah Hall died on June 22, 1830, the old tavern became a boardinghouse.[33]

[31] Elizabeth P. Nowers, *Portsmouth Women: Madams and Matriarchs Who Shaped New Hampshire's Port City.*

[32] C. A. Hazlett, *The Portsmouth Book.*

[33] Stoodley's Tavern, Walk Portsmouth Website. Retrieved June 14, 2017, from walkportsmouth.blogspot.com website: <http://walkportsmouth.blogspot.com/2011/12/stoodleys-tavern.html>

Chapter Four

Taverns, Inns and Hotels from Kittery to Portland

These taverns along the King's Highway provided entertainment, refreshment, sustenance, and lodging to travelers as well as local citizens. Stage travel was an unwanted ordeal but at times, necessary. With Maine having poorer roads than Massachusetts or New Hampshire, it became incumbent on the tavern keepers in Maine to offer as much recovery as possible. Up until about 1800, ferries across water impediments were more common than bridges, especially in the case of longer crossings.

Rice's Tavern—Kittery[1]

[1] Hubert G. Ripley, *New England Inns and Taverns*, The Monograph Series Number 6, Volume XVIII.

Jefferds' Tavern— Wells[2]

While the capital of the new state of Maine remained in Portland, transportation from Kittery to Portland played an important role in getting legislators and business men from the southernmost region of Maine to the capital and business hub of the state.

> This estate was granted by Sir Ferdinando Gorges to a family by the name of Withers, and in 1652, it was given as a marriage dowry by Thomas Wither to his daughter Mary, who married Thomas Rice. It has remained ever since in the Rice family, and seven generations lie in the family lot on the old farm. Near this lot grows a rosebush known to be over two hundred years of age. On this farm was the landing place for the ferry from Portsmouth, long known as Rice's Ferry; and through the woods passed the old road to Portland. Part of this road has been always been called Love Lane because of its beauty and seclusion, though now but few traces of either remain. The old Rice mansion stands near the ferry.[3]

The stagecoach days made lively times in Kennebunk. The first coach was owned by Josiah Paine in 1810, and went from Portland to Boston in two days. This being a success, a company was formed as early as 1820.

[2] Library of Congress HABS ME,16-WEL,3–1.

[3] Sarah Haven Foster, *The Portsmouth Guide Book, 1896.*

Joseph Barnard's Tavern—Kennebunk[4]

> The Portland Stage Company was incorporated in 1824. Kennebunk was half way between Portland and Portsmouth and the east and west bound coaches crossed here at noon, stopping at Jefferds' Tavern (Wells) for dinner and to change horses. Nathaniel M. Towle was then the landlord. Mr. Remich says in his history that twice each secular day, both from the east and west, coaches drawn by four horses rolled along the streets of the village, often followed by one, two, and occasionally three extras all loaded to the utmost capacity of the vehicle. As the horses were changed here, much stable room and blacksmith work was required. I heard Edw. P. Burnham say in an address here that one could go to Boston by the mail stage in 14 hours. The last stage from Portland to Portsmouth was in 1843.[5] Simon Jefferds opened his tavern after 1761.

In his book *The History of Wells and Kennebunk*, Edward E. Bourne, names the many taverns in that neighborhood.

> The people were establishing themselves on the seacoast at the eastward, and travel was rapidly on the increase. This was still on horseback, and men were accustomed to move slowly. Fast horses were not the hobby of this period. Taverns were now established all along the road. Pelatiah Littlefield kept one where John S. Littlefield now lives; James Littlefield near Wells Corner; Simon Jefferds where the late Samuel M. Jefferds lived; Capt. Nathaniel Kimball where Edward Haney now lives, and

[4] Courtesy McInnis Auctions.

[5] George A. Gilpatric, *The Village of Kennebunk, Maine.*

two miles beyond, in Arundel, was the Patten House. The main road at Cole's Corner turned to the right and also to the left, the right branch tending down near the sea, and the left over Cole's Hill through Harrysickett, both again coming together near the bridge over Mousam River. In 1783, and many years afterward, Cole kept a public house at the Corner. In 1800, Barnard and Howard kept public houses in Kennebunk, the former where Daniel Curtis now lives, and the latter in the old Michael Wise house. The Barnard house was continued as a tavern till after the time when this history [book] ends. Joseph Barnard carried the mail on horseback in his saddle bags until 1787, when he commenced carrying passengers, as related in Smith and Dean's Journal: 'This year the first attempt was made to carry passengers and the mail in a carriage from Boston to Portland. Joseph Barnard, the old mail carrier, got up a two horse wagon in January and put forth a most attractive advertisement, stating that he should leave Motley's tavern (Boston) every Saturday morning, arrive in Portsmouth on Monday, and leaving Portsmouth on Tuesday, arrive in Portland on Thursday. 'Those ladies and gentlemen who choose the expeditious, cheap, and commodious way of stage traveling will please to lodge their names with Mr. Motley.' 'Price for one person's passage the whole distance, twenty shillings.'

As we know, there were many small stage lines vying for a chance to be the big winner in the field of public transportation. Many tavern owners threw their hat into the ring, and for some it paid off. No doubt the lines that were linked by ownership to the taverns had the best shot at success, and several did well. Joseph Barnard passed away in August of 1817, at the age of 69.[6]

Built in 1778 by Captain Seth Spring, a veteran of the Battle of Bunker Hill, stages stopped here on the road from Boston to Portland. Captain Spring kept a tavern in this house. He also conducted a lumber business on Spring's Island and built a saw mill there. It is understood that the tavern keeper and General Lafayette were very friendly and that the tavern was named for Lafayette. When. Lafayette came to Maine in 1825, he stayed at the tavern.

Cleaves Hotel-Saco

Another stopping place for General Lafayett,e on his trip of 1825, was Cleaves' Hotel in Saco.[7] John Cleaves had run his tavern and hotel from at least 1803 until his death in 1824.[8] When the Portland Stage Company introduced its route from Portland to Saco, Cleaves Hotel was the destination, and in turn the starting point for the return trip.[9] Just as so many others associated with the taverns and stage coaching, Cleaves was a consummate businessman and citizen with a penchant for civic involvement. In 1796, he formed a corporation for the purpose of building a sluiceway from Dennett's landing on the Saco River to Pepperell's Wharf.[10] By 1812,

6 Weekly Visitor (Kennebunk, Maine) August 23, 1817.

7 New England Historical Society, *Flashback Photo: Lafayette Returns to America.*

8 Gazette (Portland, Maine) May 2, 1803 p4.

9 Maine Palladium (Saco, Maine) August 16, 1826.

10 Eastern Herald (Portland, Maine) September 7, 1796 p1.

Capt. Seth Spring's Tavern—Biddeford[11]

he was Postmaster for the town.[12] Cleaves lost his wife, Eunice, at age 65 on October 5, 1824, and passed away himself, four days later at age 68.[13]

Colonel Charles Hayes took over the tavern operation on Cleaves death and ran it until 1830, when it was auctioned off. A short description appeared in the auction notice. The house had seven rooms on the first floor and well finished chambers on the second. The third floor contained two large rooms. The kitchen and wood house were separate buildings. Also, there was a never failing well, and across the street, a lot of land containing a large stable and

[11] Collections of Dyer Library / Saco Museum Maine Memory Network Item 51586.

[12] Eastern Argus (Portland, Maine) January2, 1812 p4.

[13] Eastern Argus (Porland, Maine) October 12,1824 p2.

Alice Greele's Tavern[14]

shed. "The stand for a public house, is probably one of the best in the State, being situated on the Eastern Great Mail Route from Portsmouth to Portland."[15]

By 1899, the Cleaves Hotel and its associated clock tower had been moved to Old Orchard Beach to become part of the Velvet Hotel. It was destroyed by fire in 1907.[16]

Portland's Hotels and Taverns Eighteenth Century

Three well-known taverns existed in the Portland area before and during the Revolutionary War. These places were not especially associated with the stage lines, but deserve a firm place in the history of Portland. Two, Marston's and Broad's, continued throughout the earliest days of staging, which occurred in the late eighteenth century.

The first, located at then 300 Congress Street, was Widow Alice Greele's Tavern, a popular gathering place in Portland from 1735 to 1775. She provided home-cooked meals, lodging, and entertainment. As tensions grew between the colonists and England, the tavern became a location for heated discussions. Greele gained heroic status when she defended her tavern against the flames of a British attack in 1776. Because her building was one of the few public buildings to survive, it was used for the County Court sessions until 1787, when a new court

[14] Collections of Maine Historical Society Drawing by Charles Goodhue.

[15] Maine Palladium (Saco, Maine) February 17, 1830 p3.

[16] Maine Memory Network, Item 26573.

Marston's Tavern[17]

house was built. The building was a landmark for more than a hundred years.[18]

Willis, in his *History of Portland* tells us:

> It was common for clubs and social parties to meet at the tavern in those days, and Mrs. Greele's on Congress street was a place of most fashionable resort, both for old and young wags, before as well as after the revolution. It was the Eastcheap of Portland, and was as famous for baked beans as the 'boar's head' was for sack, although we would by no means compare honest dame Greele with the more celebrated though less deserving hostess of Flagstaff and Poins. Some people are now living on whose heads the frost of age have extinguished the fires of youth, who love to recur to the amusing scenes and incidents associated with that house.

Alice Greele kept the tavern for about 30 years, and after her death it came into her daughter Mary's hands, who sold it to James Tyler and Cotton Brown Brooks on November 19, 1802. It was afterwards a tenement house occupied by four families, and in 1846, it was cut in half, hauled to Ingraham Court, off Washington street, and burned in the fire of 1866.

Alice Greele was a nurse to the smallpox patients in 1761, and the town voted her $4.00 per week for services. She probably died about 1795.[19]

Marston's tavern was a two storied, hipped roof building with dormer windows, and stood in Monument Square where the stores occupied by George E. Thompson and Thomas Merrill, now stand [1897.] The building was originally of one story, but was probably altered to two stories by John Marston before the war. The stable and sheds are still standing in

[17] Collections of Maine Historical Society, Earl G. Shettleworth, Jr. and William David Barry, *Mr. Goodhue Remembers Portland*, Charles Quincy Goodhue Drawing.

[18] Nathan Goold, *Alice Greele's Tavern.*

[19] Nathan Goold, *Alice Greele's Tavern.*

the rear. The tavern building was moved to State Street in 1834, and is now standing on the southwest side, near York Street, but the roof has been changed. When it was moved one of Mowat's shot was found in the chimney, which then stood in the center of the house.

John Marston bought this tavern in 1762, from Robert Millions and his wife Mary, who was a daughter of Thomas Bolton of Windham. Marston was an inn holder then, and kept this tavern until his death about 1770. He was succeeded by his wife Susannah, assisted probably by her son Bracket, until 1779, when he became the landlord. In 1780, his brother Daniel succeeded him. Brackett Marston's children sold the tavern in 1795, to Caleb and Eunice Graffam. He was a soldier from Windham and they had probably kept it sometime then as the Columbian Tavern. Caleb Graffam was a post rider to Hallowell and Wiscasset, having commenced in 1791. He sold the tavern to Josiah Paine, also a soldier and post rider, in 1810. Thomas Folsom kept it in 1812, and 1815, and the name remained the same. Then came Timothy Boston, after him, Israel Waterhouse in 1823, and the last landlord who kept it until 1834, when it was moved.

This tavern's historical associations are with the year 1775, as it was to this house that Colonel Samuel Thompson's company carried Captain Mowat when he was captured on Munjoy Hill, In May, 1775; and it was here that Captain Wentworth Stuart and his men carried the five men, the crew of Coulson's boat, who were captured by Captain Samuel Noyes and his company, at the mouth of the Presumpscot River, June 22 of that year.[20]

Captain Mowat was the British commander who shelled and burned Portland in 1775.

Another interesting tavern was in Stroudwater, and was constructed and run by Thaddeus Broad.

> Broad's tavern became well known for its gaiety, its balls, as the seat of local fairs and the focal point of the muster gatherings.
>
> Traditionally, the military company assembled in front of the Tate house. The squad, never more than 25, went through drill exercises and then, two a-breast, marched to Broad's tavern with a tag along of village boys to the rear.
>
> Before entering Broad's, the men stacked arms, ready to go in for refreshment. As they emerged they again took up arms and marched to Brewers Tavern where the process was repeated. The young town boys were never admitted to the taverns. Eventually the squad returned to the drilling ground, those who could. It was to the concern of both family and friends that the men got home!
>
> Four months before Thaddeus died in 1824, his wife required a visit from the local doctor at night, who made a charge of 50¢ for senna, camphor, and

[20] Maine Historical Society, *Proceedings 1897.*

Broad Tavern–Stroudwater[21]

> opiates. On June 9, Thaddeus was dead and the charge for medical attention of 18 visits for his last illness was $9.50. Over half of the doctor's fee was paid by barter.

In 1947, the Portland Press Herald reported that the historic Broad's tavern at Stroudwater was to be demolished and that a stagecoach era landmark was to go to make way for airport development.

> One of Portland's landmarks of the stagecoach era will be pushed into oblivion soon by two popular symbols of modern progress, the airplane and the automobile.
>
> Historic 165-year-old Broad's tavern at 143 Westbrook Street will be demolished by the city to clear aerial approaches to the municipal airport and provide an additional plane and automobile parking space
>
> Built by Thaddeus Broad who came here from Massachusetts in January, 1771, the old tavern was credited by the Deering News in 1898 as being 'for 50 years a place where more really good, first class sprees were indulged in than all other places in or around Portland.'

[21] Collections of Maine Historical Society Maine Memory Network Item 40332.

When Thaddeus married Lucy Skillins of Long Creek, Cape Elizabeth, he leased the Stroudwater property from Isaac Winslow, a son-in-law of the early settler, Samuel Waldo. The first site of the tavern was believed to be on the saltbox farm owned by Colonel Thomas Westbrook. A log structure was replaced by the present frame building on the same site in 1782 and soon became a famous stage stop between Portland, York, Portsmouth, and Boston.

A shrewd merchant, Thaddeus used a large elm, still standing, as a novel outdoor drinking place. Steps were built up to where the branches began to spread and a wide platform with chairs and tables encircled the trunk. Thirsty travelers would hitch their mounts to a picket fence surrounding the base, ascend to the cool bower and according to the Portland Daily Press of 1880 be served from a 'great basket with glasses set in its periphery and a decanter of an old French brandy in the center.'

The elm became known as the Bar-room Tree, but later was named the Lafayette elm when the famous French general paused there to slake his thirst on his way to Portland in 1825.

At the height of its popularity a two-story addition joined the main house containing a hall, dining room and bowling alley. Out buildings consisted

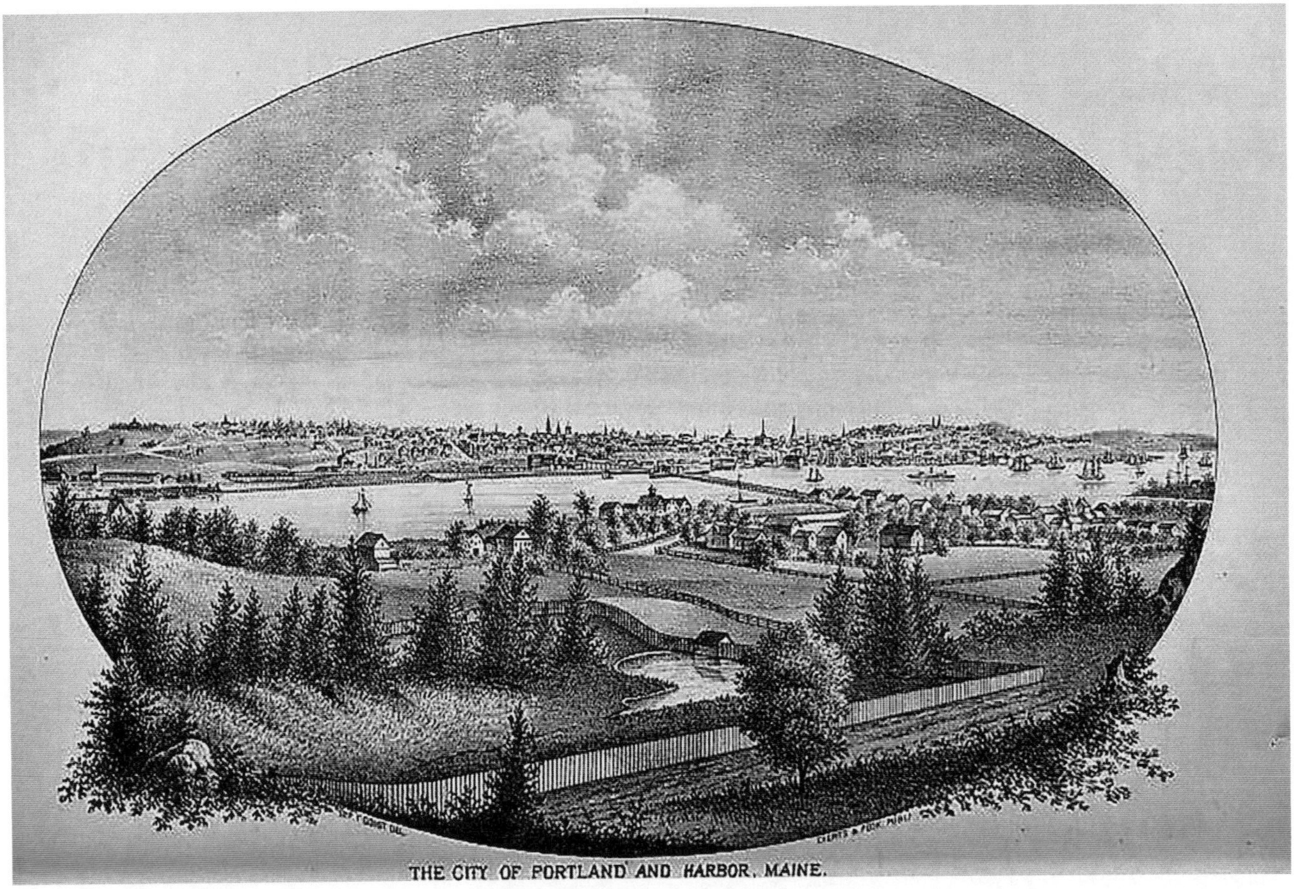

Nineteenth Century Portland[22]

[22] W. W. Clayton, *History of Cumberland County.*

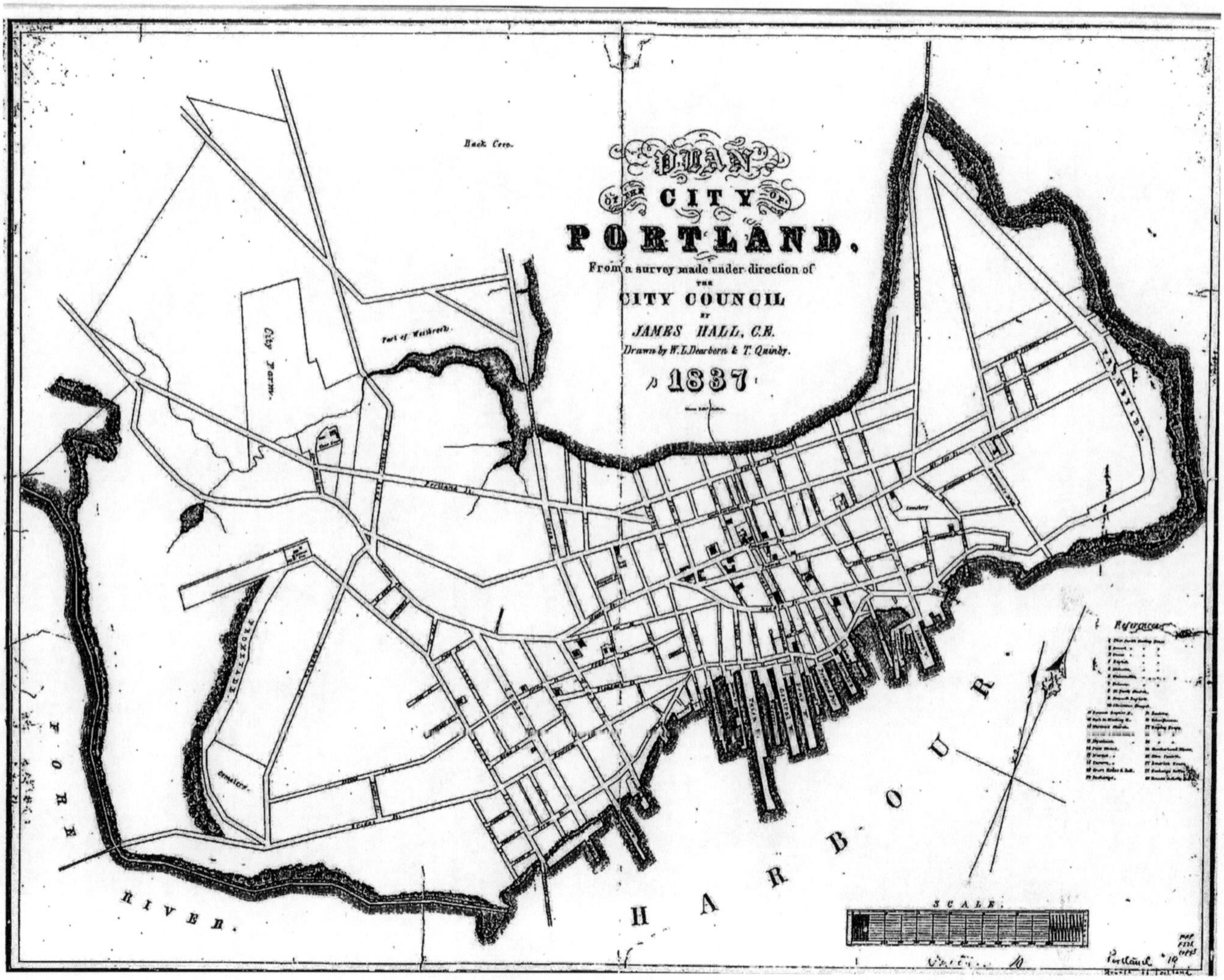

Map of Portland 1837[23]

of a shed, carriage house and immense barn.

Silas Broad, a son, succeeded as host of the tavern until the close of stage travel in 1830[?] when the elevated bar was dismantled and the home tightly closed for 50 years, but reopened in June 1880 by Miss Ann Broad, a descendant, for visitation by the public. The house was described as containing twelve large, low studded rooms, the dining-room running the width of the house and lighted by two long windows.

The Portland City Guide recounts a high point in the operation of the tavern where card games went on in the tree platform until early hours. One Saturday night a player mischievously suggested setting aside winnings made during the hours of the Sabbath and presenting the 'tainted money' to parson Caleb Bradley who lived nearby. The group thumped on the Parsons door at 4:00 AM and explained to the preacher that in as

[23] Collections of Maine Historical Society Maine Memory Network Item 4175.

American House Portland-on Fire[24]

> much as the money was won gambling he should please accept it and departed in haste to escape his wrath but the parson shouted 'Why didn't you play longer boys? The devil's money will buy just as many Bibles as God's money.'

Three taverns/inns sprang up in Portland in response to the stagecoach traffic. These were the American Hotel, Cumberland Hotel, and the Elm House, all near City Hall. The latter was known early on as Thomas S. Bowles' Stage House. Bowles ran the house for only three years, at which time it was taken over by John Appleton. Soon the Elm House was the headquarters and destination for every stage trip starting or ending in Portland.

Of the other two Inns, the Cumberland was far more elegant than the American, but the American offered better rates for business travelers.

The American House began as Whittier's Tavern in 1793, and became the Merchants' Hotel in October of 1834. In January of the following year, Joseph B. Wright and Horace Fabyan advertised the Union Hotel as the name of the establishment. About one year later it

[24] Collections of Maine Historical Society Vintage Maine Images Item 16865 *(Original in Color).*

became the America House. A fire started in the stable of the hotel on February 13, 1852, and spread to the surrounding buildings. The entire block from Elm Street to Preble was destroyed. A complete account was published in an Augusta newspaper.

Destructive Fire in Portland

One of the most fierce and uncontrollable fires with which our city was ever visited, broke out on Thursday evening about 10:00 in the stables attached to and in the rear of the American House, on Congress street, kept by Mr. Jose. The wind was blowing very strongly from the west, and the alarm was hardly given before the flame spouted up to the sky, with a most appalling energy, from the great mass of inflammable material there gathered. The two stables in the rear of that large house were soon an entire mass of flames, filling the air with coruscations of burning cinders, as magnificent to look upon as a continuous shower of falling stars.

The stables were soon destroyed, and in them perished some 25 to 30 horses, many of them of much value. In the American House stable were about 40 horses, 30 of which are supposed to have been burnt. Among them the beautiful horses belonging to Langley express, A. Witman's horse, Dr. Swett's, Richard's the Baker, and five belonging to an Ammi Ripley, of South Abington, Massachusetts, for one of which he had been offered $500 but the day before. He also lost two sleighs and harnesses. The stable next to the west was occupied by Graffam, the Miller and from this the horses got out safely by Preble Street; the stable east was occupied by O. C. Frost, Heckman, and his horses were got out by Elm Street.

The fire soon communicated to the American House, an immense structure of brick. It had but recently been fitted up and refurbished from top to bottom with new furniture. The house was full of permanent borders and travelers, and contained much valuable property. Considerable of the property was saved. For two hours it had full sway, uncontrolled. Human power could not cope with it. It was a most sublime spectacle. Occasionally portions of the walls would fall, with a terrific crash.

From the house, the old Burnham house next to it east, on the corner of Elm street, took fire but before fully ignited a portion of the huge wall of the American House fell on its roof and crushed it nearly in. This was soon demolished by fire and firemen. On the western side, on Congress street, the five wooden stores to the corner of Preble street were destroyed, thus sweeping the entire row, from Elm to Preble street, on the west and east side. The corner one was occupied by Samuel Sawyer as a grocery, insured about $1400; the next by Rufus Stanley as a refreshment room and confectionery; the next we do not learn who was the tenant; the next by Nickerson as an eating house, owned by Haggett and Allen, the last by Armstrong and Hudson. There was also a barbershop kept by John S. Libby and

Burnell's Oyster room in the basement of the hotel. Here the fire, after two hours uncontrollable sway, ceased to be dangerous the cinder's flew in sheets over the city but no danger arose from them except in igniting the gutters of the City Hall opposite, which were soon put out.

The hotel was owned by the heirs of S. Whittier and James Deering, and was insured for $7500, $5000 at Pope's, residue at other offices. Mr. Jose's furniture was insured for $5000 dollars at the two offices of which J. W. Munger is agent. The Burnham house was owned by Mr. Clapp, and occupied by O. C. Frost, no insurance on house or furniture. The furniture was mostly got out. The tier of wooden stores westerly, were all but one owned by the Deering heirs.

The origin of the fire points strongly to incendiarism. Alex Stevenson, the hack man, says he was in his office, when he heard a man spring out of a window and escape in haste, and in a minute the flames broke out.

The Advertiser says:

'The value of the property destroyed by the fire was not far from $35,000 on which the whole amount of insurance was about $23,000.'

The American hotel, which was wholly destroyed, was an extensive brick building with a granite basement, five stories in height exclusive of the attic story. Its valuation, (land included) on the city tax books, is put down at $26,000 and it rented for $1600 per annum, exclusive of the stores in the basement, and also of repairs and fittings.

The Journal learns from Mr. Carpenter of Carpenter and Company's Express, that while workmen were engaged in removing the rubbish from the site of the hotel on Saturday morning, the remains of two bodies were brought to light. Of course, they could not be identified from anything about them, but circumstances make it almost a certainty that one of the unfortunates was Mr. Joseph Linscott, of Brunswick. Mr. Linscott had been engaged in cutting ship-timber, in the south, and arrived at the American House on Thursday night on his way home. On retiring, he remarked that he had been without sleep several nights, and was much fatigued. It is supposed he slept so soundly as not to be awakened by the alarm, and so perished. No probable conjecture has been formed as to what other person had perished in the fire, as it was not decided whether the bones discovered with those of a male or female. The Argus of Monday states that on Saturday there were rumors of other bodies being found, but it proved to be parts of the one supposed to be Linscott's. Nothing has been found to show that more than one perished.[25]

Built at Hay Market (now Monument) Square in 1803, the Cumberland Hotel was originally called the Washington Hotel; then by 1829,[26] the Portland House; and on April 25, 1835, the

[25] Maine Farmer (Augusta, Maine) February 19, 1852.

[26] Eastern Argus (Portland, Maine) January 9, 1829 p3.

Cumberland House—Portland[27]

Cumberland House.[28] By 1840, it was named the United States Hotel, and even received President James K. Polk in 1846. It was remodeled in the early 1870s when a fifth floor was added, and closed in 1900.

While owned by several landlords over the years, it would seem that the hotel manager position changed much more frequently. Also, refurbishing occurred almost as often as management changed.

A vivid description was printed two days after the opening:

Cumberland House: This commodious and richly furnished house, which was opened on Wednesday last, for the reception of travelers, is thus noticed by a correspondent of the Boston Atlas:

Travelers, whether for business, health or pleasure will be gratified to find on their arrival in Portland, that it can now boast of what has long been a desideratum, a magnificent hotel. The Cumberland House, which has been recently opened, is deservedly entitled to a public patronage, from the great and meritorious efforts which had been made by the proprietor and

[27] Portland Maine History 1796 to Present.

[28] Eastern Argus (Portland, Maine) April 25, 1835 p2.

landlord to render it worthy of support. It is pleasantly situated between and at the head of Congress, Maine and Federal streets. The edifice forms a quadrangle, which when entirely completed, will be nearly 100 feet in extent on each of the sides. The chief front is south, and opposite the City Hall. The entrance to the west wing is on Congress street. The doorways are ornamented by Doric porticos and an iron veranda runs around the eastern and southeastern sides, which is entered from the parlor windows of the first story, the basement being divided into stores and offices of various kinds.

There is a spacious dining room and drawing room, with several parlors for gentlemen, in the eastern division of the hotel, and like accommodations, especially and tastefully prepared for ladies and family parties in the western, which has a separate and exclusive entrance.

There are eighteen parlors and fifty-seven bedrooms, now furnished and when the whole building will have been completed, which will be the case in the course of the summer, the latter number will be doubled and the former nearly so.

The office, or bar-room has been beautifully fitted up, with a marble counter, and all the conveniences which such a spacious establishment requires.

The dining rooms and passages are covered with superb painted floor cloths from Perkins celebrated manufactory in Roxbury, and all the other apartments with woolen carpets. The whole house is splendidly furnished at an expense of over $15,000.

The tables are luxuriously supplied with the most delicate viands, and all the choice products of the sea and land, and which are served up in the best manner, by a thoroughly educated artist, who is in charge of the culinary department; and his ample battery affords all the facilities which are requisite for the full display of his consummate skill in a science so interesting to the traveler, especially to feed the one of the disciples, or even a real amphictyony of gastronomy.

The cellar is enriched with the trophies of all the most celebrated vineyards of Europe where the connoisseur can eloquently expatiate on the names, ages, flavor, and aroma of his favorite lines, while the man of temperance has an excellent opportunity of testing the sternness of his abstinent virtues, for it is only when in the way of temptation that one's perfectibility is to be verified.

The host appears to be well qualified for his station and most desirous of giving entire satisfaction, and the numerous well drilled attendants are emulous to administer to the wants of his guest.

If at the Cumberland House, gentleman, as well as ladies, cannot be comfortable, a most comprehensive and significant

Elm House and Tavern—Portland[29]

> word, peculiar to the English language, why, then they must be predetermined to be miserable everywhere, and as Major Jack Downing observes, 'where's the use of doing your best to please the people, if they won't be satisfied with nothing? You might as well attempt to bleach snow, or harrow in hail for hay seed, as to try to suit a willful man or a notional woman. You can't make 'em smile any more than a basket of chips, do which you will.'[30]

The earliest notification of the existence of the Elm House occurred in 1825 when Thomas S. Bowles announced that he "has taken over that new and elegant brick house on the corner of Federal and Temple streets, lately erected by Hale and Waterhouse, where he solicits the patronage of that part of the community who have long experienced the want of a comfortable, convenient, and well-furnished stage house. They may be assured that no expense or labor shall be spared to give satisfaction to all those who may please to call on him. The stage books are kept at the above house as usual and every intention shall be paid to stage business for the information of the traveler. A few genteel borders can be accommodated at the above house."[31]

A year earlier, Abraham Colby assumed management of the "well known stand in Portland formerly kept by Israel Waterhouse."[32]

[29] Advertisement in *A Business Directory of the Subscribers to the New Map of Maine with a Brief History and Description of the State Prepared by William Willis.*

[30] Portland Daily Advertister (Portland, Maine) April 27, 1835 p2.

[31] Eastern Argus (Portland, Maine) December 13, 1825 p3.

[32] Independent Statesman (Portland, Maine) January 10, 1824 p4.

Advertising Flyer for Hack Service[33]

We will learn more about Israel Waterhouse and Samuel Hale, founders of the Maine Stage Company, in Part Two. They too understood that owning a tavern and hotel was the best means for making a living with stage coaching. Bowles was succeeded by J. W. Appleton around 1828, and the Elm House became the Appleton Hotel.[34] During his tenure, Appleton succeeded in getting all the stage lines which came to Portland to start or terminate their routes at his hotel.[35]

33 Collections of the Maine Historical Society Maine Memory Network Item 11224.

34 Portland Advertiser (Portland, Maine) September 1, 1829 p3.

35 Maine Memory Network, Portland Hotels. Retrieved June 14, 2017, from MaineMemory.net website: <http://www.mainememory.net>.

In November of 1833, S. Haskell, formerly of the Lafayette Hotel in Boston, took over the Elm House.[36] Three years later he advertised that he had the stage books for the Eastern and Western stages. Also, stages to the White Mountains, Waterford, Paris and Central Harbor started from the Elm House.[37] Haskell only kept it for five years as Mr. O. P. Thorp announced in November of 1841, that it was reopening after a thorough cleaning, painting and repair to bring it up to the standard of "modern hotels." He also claimed that it now had new bedding, bedsteads, and furniture throughout. His assistants were said to have "accommodating dispositions."[38]

After Israel Waterhouse died in 1842, his two fifths part of the Elm House was offered for sale for $1400, the amount required to cover his debts. Others carried on the Elm House until the great Portland fire of 1866. The house was damaged, but afterwards, was rebuilt and became the headquarters of Kendall and Whitney, a distributor of agricultural machinery.[39]

Although we do not know when this poster was published, it does remind us that local transportation by hack, or taxi, prevailed even after the stagecoach had been mostly replaced by the railroad and steamship.

The availability of these three fine hotels, and at least a half dozen others, was a direct result of Portland finding its place on the map. Thanks to the stagecoach and steamship businesses, and of course the railroad, people were beginning to see Portland as a "go to" destination. The Portland Advertiser did a column on the Portland Hotels in 1851:

Portland as It Is.

The natural attractions of our city, are at length beginning to be appreciated abroad, at least so it would seem from the encomiums heaped upon us by the crowd of leisure writers, who have 'brought us to light' in their journeyings to the White Mountains. It is no news to our citizens that Portland abounds with the 'female loveliness' and intelligence; but many in our midst have become so accustomed to 'noble scenery,' that they seldom give it a look or a thought.

There is a variety and beauty in the views commanded by the two broad 'promenades' that encircle the city, views of the far off mountains, of the bay with its evergreen islands and romantic shores, of the broader ocean, of the extended line of seacoast, of the harbor and shipping, of the wooded hills and winding waters inland, which no stranger of taste can contemplate without being surprised and delighted. Then, our affinity to Cape Elizabeth with its pleasant watering places, and its rocky headlands, hard sand beaches, and rural retreats, any of which can be reached by an hour's ride at most, our proximity to the retired coves and inlets, and forest shades of islands, and to the excellent fishing grounds of this section, quite as

36 Eastern Argus (Portland, Maine) November 25, 1833 p2.

37 Portland Advertiser (Portland, Maine) April 22, 1836.

38 Ibid November 30, 1841 p3.

39 Maine Memory Network, Portland Hotels. Retrieved June 14, 2017, from MaineMemory.net website: <http://www.mainememory.net>.

First the New England House and then the City Hotel[40]

easily accessible by steamer or sail boat, and above all, our annexation to the White Mountains by railroad, to the Alps of America, which can be reached in three or 4 hours, now combine to make Portland one of the most desirable places of summer resort in New England. This, as we have before intimated, is beginning to be appreciated by pleasure and health seekers from abroad, and in proof of which we have only to point to the fact that our hotels, of which we have nearly a dozen, have been crowded for the last three or four weeks; so that some of our enterprising citizens begin to talk of the necessity of a new public house.

There can be little doubt that we need a new hotel, one whose architectural proportions and style of fitting shall be in keeping with the growing importance and popularity of our city, a first class hotel, which, besides affording a pleasant home for business, shall offer inducements to people of wealth and leisure to spend a portion of their time with us and we hope that capital may be found, before another year passes to build one that

[40] Collections of Maine Historical Society Maine Memory Network Item 20585.

shall be second to none in New England, in regard to appointments, taste and keeping.

We would not have travelers suppose from these remarks, that they cannot be well provided for here. On the contrary, we have a large share of hotels, compared with the number of our population, some of which are kept in a very superior manner. To enumerate, there is:

> The United States Hotel, at the junction of Congress and Federal streets.
>
> The Elm House, corner of Federal and Temple streets.
>
> The American, 259 Congress street.
>
> The Casco Couse, Middle street opposite head of Plum street.
>
> The Union Hotel, on Fore street, near corner of Exchange street.
>
> The Commercial House corner of Fore and Willow streets.
>
> The Atlantic House, foot of India street.
>
> The Depot Hotel, at Canal street, near foot of State street.
>
> The Adams house, corner of State and York Streets[41]

The City Hotel was owned by J.W. Robinson & Son, and located at the corner of Congress and Green streets, in Portland. Green Street later became Forest Avenue.

The hotel was built in 1830, and was originally called the New England House. It was demolished in 1894 and the Congress Square Hotel was built at the site.

[41] Portland Advertiser (Porland, Maine) September 2, 1851 p3.

Chapter Five

Personal Stories Boston to Portland

A Visit to Broad's Tavern

Herbert Milton Sylvester in his book *Olde Cascoe*, recounts a trip to Broad's Tavern in Stroudwater, Maine sometime before 1909. By then the house was no longer a tavern but rather the simple residence of 71-year-old Miss Amira Ann Broad, a descendent of the original owner, Thaddeus Broad.

> Like all old roads, this artery of travel into the inland, from which the little hamlet of Stroudwater draws its nourishment, holds many a surprise for the wayfarer who follows its narrow trend for the first time. From the top of Stroudwater Hill one sees a group of stately elms; and within the gray shadows of their shapely domes, doubly conspicuous from their height and massive proportions, and their isolation in the midst of the rolling farmlands, a cluster of dark roofs of ancient aspect that have upheld the honor of their builder these hundred years and more. This place was known as Broad's Tavern over a century ago, one of the famous hostelries along the ancient coach road to Boston, past which the lumbering coaches went on their way, to or from, old Falmouth Town twice a week; which was something in the way of traveling accommodations for those times; considering the primitive condition of the roads, that gave the traveler a shaking-up that lasted several days. To be exact, this superior service dates from 1760, before which time the mails were very irregular; mail matter not being dispatched until enough had accumulated to pay the carrier, who came, and went with it, on foot, carrying the mail bag on his back. After a time, horses, and the more

convenient saddle bags were used; but the mail came and went as leisurely as ever. A schedule of arrival and departure was a thing unthought of. A case in point is an anecdote told by Willis, of a Falmouth gentleman who, by stress of business was obliged to make the trip to Boston, no mean undertaking then, who had waited several days for his mail, but the mail carrier did not come. Impatient to be off, the gentleman began his journey. He met Barnard, the carrier, in Saco Woods, where the mail was deliberately opened by the roadside and the wished-for communication delivered. Barnard's honesty must have been of the proverbial 'Down-East ' sort.

The deep ruts that once turned into the ample tavern yard are gone; likewise, the big sign that swung to and fro in the shadow of the big elm across the road by the barn. The only suggestion of the former, is a narrow footpath made by the house folk in their commonplace goings and comings; while only the gray, weather-stained post, leans out over the highway to still remind the traveler of its ancient occupation, as lonely and neglected now, as it is barren of its old-time importance. I do not imagine that the old Broad Tavern was so much different from that famous Wayside Inn frequented by Longfellow.

Better still, as I turn into the spacious yard, under the great elms, that from the hilltop, look so much like a great green dome, I see an old estate with hardly a single sign of decay about it, unless it be the sagging ridge polls of the horse sheds, that extend down the road from the barns that stand as staunch, as though 100 winters had not hurled their sleet and drifting snows, and January rains against their moss patched gables. This old tavern is one of the best-preserved houses here about; and when it was my good fortune to see all there was to see about the old place, it possessed a store of antiquated things that would turn the head of any bric-a-brac hunter.

This was comparatively a new country when Thaddeus Broad came hither, more than a century ago, to build his cabin and with his good wife Lucy, settle down beside the old trail, which was soon to be the great thoroughfare between the more important settlements of three states. All of the worldly chattels of the elder landlord of the Stroudwater Inn, coming hither from the older and more populous Massachusetts Colony, were carried in an ample handkerchief. Here at the edge of Stroudwater, he began a humble enough career, with his saw, broad axe and hammer, to get together a shelter which the wayfarer of those days was willing to accept when overtaken on his journey by the nightfall; or a storm that swept Inland from the sea, scouring the sand-dunes and marshes. This entertainment grew to be a custom. The little house on the Stroudwater road was enlarged into a commodious tavern. The barns were built, and new lands were cleared for grass and grain with which to fill them.

A sign was swung to the winds, and the criticism of the traveler.

One can see a bit of the old black sign, once ambitious enough, at the old place. I found it in the Broad tool house along with the last one that swung from the Broad gable. On one side of this old relic was a painting of the frigate Constitution under way, with all sails set. On the reverse were depicted a Continental soldier and a red coat in belligerent attitude. Appropriate and patriotic mottoes might be read on either side of the old sign; while across the lower panel were printed the name of the tavern keeper and the date of the tavern 'house warming.'

In 1834, the son, Silas, replaced this ancient and much shattered symbol of his father's hospitality with one which resembled a huge bunch of grapes painted a bright yellow against a wooden background of vine-leaf deftly carved at the edge. This hung from a large wooden hand until long after the railroad was opened eastward from Portsmouth to Portland, which soon perceptibly affected the traveler over the old Boston road, and likewise revenue of the Broad hostelry.

For years Silas broad kept open house; and with him passed away the routine of tavern keeping but not the flavor of old romance that was peculiarly appurtenant to the Broad acres and savory chimney smokes, nor the legend of its hospitality, which hospitality indeed is to this day generously dispensed in private life, under the same old roof-tree, by the last of the line, Miss Almira Ann Broad, whose horizon line of Stroudwater woods does not by any means mark the boundary of her influence and philanthropy.

The Broad's were a hardy and toil toughened race. Lucy, the first hostess of the famous Broad's, died at the age of 105, and the present dweller at the old inn, at the ripe age of 71, possessed of the freshness and the vivacity, and ruddy health of a woman in the prime of life.

Standing on the broad flag of granite that held the approach to the gabled porch, apparently the nowadays entrance to the big house, I grasped the black iron knocker; and a strange metallic crash of sound went clattering down the hall within, up the great kitchen, to tell, with hollow voice, its message. A feeling almost of remorse stole over me at this cold-blooded invasion of what seemed a sacred precinct; for I was a stranger to the people who lived here, the direct descendants of old Colonel Broad who might reasonably be expected to resent such flagrant curiosity. But no one answering, I sent the echoes of the huge knocker flying through the house a second time. From the door opening just a bit, I caught a glimpse of a pair of mild brown eyes, with just a hint of doubt about them, peeping out between its narrow edge and the stout pine lintel. Satisfied with the preliminary survey, a sweet-faced woman with white kerchief pinned about her shapely shoulders, her hair with just a hint of silver in it, combed straight back, without a single artificial touch or garnishment to mar its simple beauty, stood within its shadow.

'I called to see the house!' I said.

'Will you walk in sir?' was the gracious response.

Over the charmed threshold, down the long hall, into the old-fashioned sitting room I went. One side of it, mostly of glass, looks out over a green slope of mowing-land extending down to the woods that partly hide the blue waters of the creek. I sit in the big rocker that was the favorite a century ago, with a sense of restfulness that makes the chat of old times and the old house delightful enough. A bright fire of seasoned birch is blazing upon the wide hearth that has burned out many crackling backlogs; and upon whose glowing coals many a mug of flip has set 'a-simmering' to serve its turn with the travelers and wags who sought the hospitality of the old tavern. Incongruous as the fire may seem, with the roof-tree overshadowed by the green tops of the elms, it was cheerful on this morning early in June, for the days preceding had been days of cold rain, leaving a feeling of deafness and chill about the old house, for all the summer sun was shining so brilliantly upon the fields and woods. It added to the pleasure to see the old hearth made young again in its glory of leaping flame.

It was a great place for winter dancing parties from town, and it is not difficult to imagine a beautiful picture of an evening at Sudbury Inn having its counterpart at this ancient rule and in Stroudwater.

In this crackling birchen flame, I see the gay town folk who have come out here for a good time, in their smart costumes; their hair white with powder, and their cheeks berouged and bepatched; the gentlemen not a whit the less stylishly gotten up; with their long queues done up in ribbons; their silk hose and velvet breeches; there embroidered waistcoat and dainty laces; their silver or gold knee buckles and pumps, waiting for some tardy exquisite who was looked upon as the leader of this jolly set; or it may be the fiddler who is belated, for a dance is nothing without a fiddle, and an old fashion fiddler to fiddle it.

But the time has come for the festivity to begin. There is a hush in this youthful hilarity that is merged in the bustle incident to the more immediate preparations for a stately minuet or a more rollicking measure still. It was a dissipation that was kept up into the wee small hours of the morning, if the chronicler of these events is to be believed, and much to the scandal of the community; for to the orthodox mind in those days, dancing was a lure of the fiend. In Provincial Falmouth dancing was prohibited by law, in places of 'public resort.' The 'quality' in town held their dances at Freeman's Tavern, an old-time hostelry of most excellent repute, and on one occasion, as early as the first year of the revolution, the dancers were indicted. Among them was Theophilus Bradbury, who afterwards became a distinguished lawyer, and with whom the distinguished Theophilus Parsons studied in after years. Bradbury appeared for the respondents with the ingenious defense that as the dancers had hired the room for the season, it became a private apartment and was not a place of 'public resort.' The court sustained the council's view of the case and the 'quality' danced to their heart's content ever after.

Along a narrow, old-fashioned mantle, so high up that I could no more than easily reach it standing, were the same old candlesticks a row which belonged to the earliest days of this inn and which gleamed as kindly and looked as gay as if they had just come from the modern manufacture of bric-a-brac; only there was the not easily describable flavor of antiquity about them which is lacking in the modern article.

The ancient brass mounted andirons, the fender and heavy tongs, and the long slender backed broad bladed shovel polished to the brilliancy of gold, keep them demure company about the broad fireplace, that with its short, chunky jambs speak of the stout heartedness and toughness of things in general when its virgin flues were first aglow with flame. What tales these quaint appurtenances of this old room could tell, with its medley of experiences of home life, that began with a hanging of its stalwart crane, the drawing of its child life, the incoming and the outgoing of its stranger guest, its episodes of roistering entertainment, and its midnight revels! What a store of precious secrets are held within the heart of its old roof-tree of pine; as sound, every timber about it, as when with big broad-axe they were hewn square, and with mallet and chisel, were fitted into a perfect roof plate, rafter and ridge pole! A square house, goodly in proportions, set upon capacious foundations, with two good stories above. It is painted white, with cool looking green blinds, to give a pleasant contrast; and from the eaves on the side toward the highway, it's sharply pitched roof runs up to a stout ridge pole with its single stout chimney amid ships; to make as steep a dissent on its rear side, keeping on down over its ell, shed like, until its low eaves overshadow the windows of the ancient kitchen. East and west, its gables look with the highway. From the horse sheds eastward, it is a delightful vista of birchen woods over the June landscape to the farthest point of the rising, old 'Black-strap,' with its wooden monument, a relic of a costal survey made in the early part of the century. Westward from the hooded doorway, with its sidelights of green glass, one sees the sun set amid the orchard tops; and that is all.

From the restful entertainment of this old room and its smoldering hearth fire, the musical speech of my gentle hostess in her suit of gray, and the June sunlight without, with the west wind blowing through the orchard and into the open windows, bringing with it a bar or two of some orchard singer's madrigal, it is but a step to the quaint staircase with its slender handrail. The shadows thicken as the garret is approached with its single window in either gable, a roomy, unfinished interior, rich in memorials of the time and a people, the simplest episodes of whose most matter of fact existence are tinged now with the color of romance.[1]

Attempt to Rob the Mail

An attempt was made on Thursday night last in Seabrook about seven miles from Newbury, to stop the eastern mail coach, by a person who sprang from the roadside, seized the reigns of the leaders and drew them out of their course. Mr. Potter, the driver, by skillful management of the reigns in giving his horses the whip, immediately succeeded in turning the horses into the road and disengaging them from the villain. A few minutes after he saw three men coming from the field into the road but he was driving with great rapidity, and they made no attempt to molest him. There were only two passengers in the stage. The mail agent at Newburyport yesterday dispatched a person to make proper inquiries. By the Portsmouth coach we learn that the persons sent from Newburyport on coming to the place where the mail coach was attempted to be stopped, found a pole line across the road, from which it was concluded that a robbery of the mail was intended. The mouth of one of the horses is severely cut by the effort made to stop the coach.[2]

A man by the name of Fifield of Stratham, was yesterday found in a barn in Seabrook, much bruised; his wounds were probably inflicted between the coach and the fence. Seabrook is seventeen miles from Dover and eight east of Newburyport.[3]

A Matter of Heredity

A descendant of one of the earlier families in Portsmouth which resided between Sagamore Creek and Great Swamp, was traveling in a stagecoach with a stranger who found that they both had the same family name. On inquiring of descent, the stranger, of somewhat high notions, said he was

1 Herbert Milton Sylvester, *Olde Cascoe* [Drawings and Text.]

2 Hallowell Gazette (Hallowell, Maine) April 1, 1818 p3.

3 Sun (Dover, New Hampshire) March 31, 1818 p3.

connected with a family of Sir David B. of Scotland. The native of Portsmouth, who thought his claims to be of aristocratic descent no less prominent, replied that he was descended from a family of Alderman.

'Family of Alderman,' said the sprig of nobility, 'why, you must be very ignorant to think that there is any such hereditary order, it is only a temporary city office, sir.'

'You are mistaken,' was the reply. 'It is an order which ranks a little higher than knighthood. My forefathers for five generations bore the insignia of their high honors. They wore the Alderman's aprons with as much honor and pride as any Knight Templar. Those aprons were no fragile silk or linen fabric, they were the pure hide, such as were used when the ark was constructed; and they date their nobility at as early a day. Perhaps you may yourself one day arrive at the honors, and then you will fully comprehend them. These aprons they wore six days in the week, and the ravages they made around Great Swamp and Sagamore Creek, are now manifest in the well cleared and productive farms of their descendants. Yes, sir, I am a regular descendant of the family of Alderman and shall never lose my aristocratic pride, but will endeavor to respect those who may be of lower rank.'

To be descendants of the family of Alderman should be the pride of American nobility. Of such was President Lincoln.[4]

Teaching Politeness

In the year 1815, there resided at the stage house in Portsmouth, a gentleman (we shall call him Mr. P.) who always took the liberty at the table of helping himself to the best that was set before him, which piece of impudence caused unpleasant feelings in some and wry faces on others, who were for sharing the food in common. Within a short time after this gentleman commenced boarding at this house, he became very notorious for his impoliteness, and passengers who arrived in the stage had other tables set for themselves wholly on account of the aforesaid gentleman. A large party from Boston arrived at this house in the spring of 1815, and met Captain N. of Newburyport, who by the way was an old sea captain, and had been on the ocean from his childhood, amid the snows of Greenland and the heat of the West Indies. He was plain in his manners, pleasant in company, and a very agreeable companion. The gentleman from Boston gave Captain N. the description of Mr. P before dinner both in person and manners. They all sat down to the table about 2:00; Captain N. Took a seat next to Mr. P. and waited with patience till the gentlemen took the first cut into his own plate as usual when Captain N. removes Mr. P's. plate towards himself and placing his empty one before him, very politely thanked him. The house resounded with laughter at Mr. P's expense. It had the desired effect. Mr. P. was never known to be guilty of a breach of politeness afterwards.[5]

[4] Charles W. Brewster, *Rambles About Portsmouth*.

[5] Portland Advertiser (Portland, Maine) August 6, 1830 p1.

Traveler in Maine

Portland, June 17, 1845.

Brother Woodbridge, -you'll hardly expect any 'pencilings by the way' quite so soon after leaving your goodly city. But having got so far 'Down-East' in five hours from Boston, and it being just 70 years from the Bunker Hill Battle day, and just 100 years since the capture of Cape Breton and two days before the great gathering of 'Odd Fellows' on Boston Common, and the presence of spectators, if spectators can be found after the embodiment of all the odd ones there are in this odd world, I thought I might as well make a sort of beginning, though it may seem a little like beginning before one has anything to say.

Now a few words about railroad travel. As this is the first time I'd ever entered the city by railroad, I may be allowed, in the absence of anything better to say, to refer to one or two other trips from Boston 'Down-East.' The first was not far from 20 years ago. I took 'the mail stage,' 2:00 AM at the Eastern Stage House, where all passengers must be the night before, as that line 'called around' for no passengers. We left at that unseasonable hour in the midst of the first snow storm of the season. Before we reached Newburyport, where we found breakfast in waiting, the snow was so deep that we could hardly tell whether we're making headway or not, except as the driver was able now and then to give us a gentle shaking by getting into a pretty decent 'rut, 'or coming in contact with a stone that happened to be on the track. Every other day the stage went over what is called 'the upper route' from Portsmouth, through Dover, South Berwick, etc.., Which was much longer than the 'lower route,' through York, etc. It so happened that our traveling day was 'the upper route' day. We jogged on from Newburyport to Portsmouth, then to Dover, (where by the way, as it was but a day or two before Thanksgiving in New Hampshire, we found a first-rate dinner,) and then onward we went and were within 38 miles of Portland at 5:00 PM, fifteen hours from Boston, when I left the stage and took a private conveyance to a town in the interior. At another time, I left Boston at the same place, in the 'Mail line,' at 5:00 PM, and arrived at Saco, fifteen miles from this place, at 7:00 the next morning; 100 miles in fourteen hours, or as we should say at that season of the year, from 'sun to sun.'

Yesterday, I left at 4:30 PM and at 9:30, 5 hours, with the detention of 25 minutes at Newburyport, for the eastern train to pass, arrived in this pleasant city. So much for improvements in traveling. But by improvements I refer not only to the saving of time and fare, but also though shakings, the fewer of which one receives, the better he feels.

Nothing special occurred in this five-hour trip, except it be that we came very near rising involuntarily upon our feet, as we emerged from underground on leaving Salem, when the whistle gave a most unexpected and tremendous outcry, and the brakes were in an instant put into operation to stop the cars. 'What is the matter?' was the inquiry; 'the draw is up,' was the answer and a most satisfactory one it was too. The vessel passed through, the draw was let down, and all was safe. I was not aware before that the iron horse could be brought to a dead stand so quick.

I sometimes hear it remarked that railroads are ruining our country. I have never been able to understand or see wherein the truth of this remark lies, and certainly never more so than at this time. I say not this as a political man, or as one that has anything invested in railroads, but I say it, taking the world as it is, and as it is to be. Who does not see that the facilities for traveling and for conveying freight, must keep pace with the increase of population, or this increase of population cannot be provided for. When something more than 200 years ago, a committee sent out from Boston to explore the country and see what the prospect was for immigrations, succeeded in getting to Stone Brook, in Weston, twelve miles from Boston, and returned, stating in the report, that they considered it unnecessary to go further, as it was not likely the country beyond would ever be settled, there was no need of railroads. And when our fathers found their way to and from their place of labor 'by spotted trees,' as I've heard my father say was the case with him, railroads were not needed. But inasmuch as the country happened to get settled a little more than twelve miles west from Boston, is there no need of rail-roads? Freight for the supply of the town or country cannot now be drawn on a hand sled once a month or once a week.[6]

An English Gentleman in North America

March, 1821

At Portland I found, a respectable boardinghouse where lodged, among other persons, were the Governor of the state, the speaker of the House of Representatives, and eight or ten of the most respectable members. There was a common table at which all ordinarily assembled; and a common sitting room, where they seemed to pass their leisure in reading the newspapers and smoking cigars. For the very first time since my arrival in America, I had actually at this boardinghouse a parlor to myself, which arose from the circumstance of its being, in the first instance, designed for my bedroom. It was a luxury indeed to feel alone, and likely to remain so, without shutting myself up in my bed chamber, in which I have lived for the last year when not in society or on the road. My hopes of retirement in my parlor, however, were soon shaken; for the landlord brought a gentleman to me, who, after conversing a few minutes, said, he was come to take me into the dining room, to introduce me to the company. He was a young lawyer, gentlemanly in his manners, and I found afterwards had been educated at Harvard college,

6 New England Puritan (Boston, Massachusetts) June 27, 1845 p1.

Cambridge. As we sat down to dinner, at one o'clock, he introduced me to most of the gentleman by name, and among others to the Secretary of the State. The rest of the company, although I doubt not intelligent and acute, I certainly should not, (at least on my first arrival in America) have guessed to be a body of legislators. The landlady presided, with Mrs. ____, the wife of the speaker, on her right and the landlord sat down towards the close of dinner, after having waited on his guests, and assisted the waiters till all of the company were helped. He was very civil and came into my room half a dozen times in the course of the evening to look at my fire, and see if I wanted anything. An English landlord could not have been more respectful and attentive. In the course of the evening, the young lawyer also paid me a second visit, with really good nature, bringing in a friend, 'lest I should be lonely.' I gave you these little incidents to show you the habits of the country. As they found me busy writing, however, they stopped only half an hour, and retired saying, that they would not interrupt me, but would attend me to any church in the morning to which I liked to go.

While in Portland I found the snow in many places two feet deep for a great distance, and perhaps fourteen inches deep where it was the thinnest. I counted twenty-two sleighs at the church door on Sunday. I saw the town under unfavorable circumstances; but it had a very respectable appearance, many of the houses being large and handsome, with extensive courts before the doors, ornamented with shrubs and grass plots. The bay and the adjacent scenery are very picturesque.

We left Portland at 5:00 AM on the 19th. The roads were so much blocked up with snow, that the mail and passengers were obliged to be carried in an open sleigh: It was very cold: the thermometer, I should think, not being above zero degrees: but the moon shone so brightly on the new fallen snow, that we should have been sorry to have missed this beautiful winter scene, by being cooped up in a close carriage. We reached Saco, fifteen miles, to breakfast, when it was determined to dispatch us in two sleighs, our unicorn equipage being found inconvenient in the snowdrifts, from having two horses a breast. James and I were put into a tandem sleigh about as large as a parlor coal box, or a little larger, the driver standing up to drive. Our two companions followed with one horse and a similar sleigh; and the way we went over the snowdrifts, the music of our bells resembling a constant concert of Jew's harps. Sometimes the bells of our companions suddenly ceased or literally 'dropped:' for, on looking behind, we used to find that their horse had disappeared, his chin resting on a snowdrift, and his countenance exhibiting a most piteous expression of helplessness. At other times our horses fell through, and it was with great difficulty we extricated them; the snow being sufficiently frozen to be of a very inconvenient consistence, although not hard enough to carry us rapidly on its surface. Our horses were sometimes prostrated three or four times in 20 yards. Once we were obliged to be cut out, and another time to have more than 20 men and several oxen to clear our way, the drifts on the road being from six to twelve feet deep. As we had excellent drivers however, who drove with great rapidity where the road would permit that, we reached Portsmouth 60 miles from Portland at 4:00, eleven hours, after an amusing and agreeable, and in some degree, adventurous ride. The cold morning was succeeded, as is often the case in this fickle climate, by a beautiful warm day; and although the road, except in the vicinity of the pine hills is rather level, the fur groves and large masses of rock often combine with the open sea, which is almost constantly insight, to form rather interesting views. The country is tolerably well settled, and we passed through several little towns; but the houses being less frequently painted than other parts of New England, have neither the same neat nor flourishing aspect. The people however seen everywhere busy and robust.

Portsmouth is a noble harbor on the Piscataqua, which is so deep that the vessels discharge along the wharf; and so rapid, that even in this winter, the severest which has been in America for at least 40 years, it's navigation has never been interrupted. A navy yard is established near the town, where 'The Congress' and other ships of war were built, and where they are now building a seventy-four-gun frigate. As the best boarding house in Portsmouth was full, I went to the Stage Inn rather a dirty scrambling tavern, where I found at breakfast the next morning, amid a motley group, one of the judges and several lawyers. The Supreme Court was to be opened early in the morning; and as it was before my hours of commercial calls, I attended to hear the jury's foreman, and the judge's charge. Both the grand and petite jury in the appearance of which I could discern no difference, seem to be composed of respectable yeomen, of about the same rank as our farmers of 300 to 500 pounds per annum. They listened with great attention as the judge read (not spoke, which greatly took from its effect,) a plain sensible charge, much to the point. The aspect of the court in general pleased me from the homely suitable appearance of those of whom it was composed; homespun clothes, with large buttons and long waists, waistcoats, with the old triangular indenture or pointed flaps, and hats with good broad brims; the absence, in fact of all affection of fashion, or awkward attempts at city spruceness. This has pleased me particularly throughout New England and forms a contrast with the style of dress which meets the eye generally in passing along the seaboard of the middle and southern states where blue coats, black waistcoats and blue pantaloons, produce a monotony far less agreeable and picturesque than the variety of dress adapted to the variant employments of the wearers. I had little opportunity of seeing the society of Portsmouth, as my stay was so short; but I met with some whose manners convinced me that I should have found a refined and polished circle there, if I had remained. From Portsmouth we reached Newburyport, where I walked down towards sunset or sundown, (as it is always called in this country,) to the mouth of the Merrimac, and had a noble view of the open sea. The roads in this part of the country are excellent, and the finger post are so like ours pointing to Salisbury, Ipswich, etc. that it was easy to imagine myself in the south of England. In most towns in New England the houses generally stand alone in a courtroom garden with lofty trees in their immediate vicinity. The inn was a large brick house in which I had a spacious bedroom as neatly furnished as at the principal inns at Bath or Chiltonham. I rose fairly early the next morning and spent half an hour in a church yard in the neighborhood, with the hope of seeing the sunrise clear out of the Atlantic, a few hours after it had risen on you all in the east; but a little invidious cliff intervened. The ocean however was beautiful; and this quiet churchyard on a foreign shore gave rise to many solemn and very interesting reflections. The 22nd was Washington's birthday, which in the principle cities, generally gives rise to public dinners and balls. Here the afternoon seemed to be made a holiday, and the young men turned out in great numbers, very nicely dressed.

On the 23rd I left Newburyport for Salem, 25 miles distant, where we arrived at noon. The surface of the ground was generally well cultivated; but I often observed immense rocks apparently growing in the fields, and evincing that the country immediately on the coast was more indebted to man than to nature for any appearance of fertility it might exhibit. Indeed, I think a great part of the road between Newburyport and Boston presents a rockier region than I ever before saw in a state of cultivation; but everything seemed to yield to the proverbial perseverance of New England. I've seen a New England clearing what appeared to me a bare rock for the sake of the narrow strips of soil in the crevices; and I could not help thinking with what a smile of contempt a Mississippi or Alabama planter would recall such a scene to his recollection, while standing with folded arms over his slaves as they hoed his rich alluvion, but both his contempt and pity would be sadly misplaced.[7]

The Capital of Maine was Portland at the time of this writing.

[7] Essex Register (Salem, Massachusetts) September 4, 1823.

State House and Exchange Tavern on State Street in Boston, Massachusetts 1801

State House in Portland, Maine 1820

State House in Augusta, Maine 1836

Merrimack Hotel and Wolfe Tavern in Newburyport, Massachusetts

Earl of Halifax Tavern in Portsmouth, New Hampshire

Stoodley Tavern in Portsmouth, New Hampshire

Parlour of the William Pitt Tavern in Portsmouth, New Hampshire

Rice Tavern and Landing in Kittery, Maine

Barnard Tavern in Kennebunk, Maine

Buxton Tavern in Walnut Hill, North Yarmouth, Maine

Peacock Tavern in Richmond, Maine

Cushman Tavern in Lisbon, Maine

Bachelder Tavern in Litchfield, Maine

Hathorne Tavern at Day's Ferry in Woolwich, Maine

The Yellow Concord Coach 1830

Lambert Tavern Sign in Falmouth, Maine

Washington Hotel Sign in Lynnfield, Massachusetts

Wallpaper Merrill Hotel in Durham, Maine

Display of Shepard Tavern Parlor in Bath, Maine

Bath Hotel in Bath, Maine

1823 First American MacAdam Road

American Stage Coach 1827 to 1828

Part Two

Introduction to Part Two

Portland to Augusta

The Statehouse in Augusta-1853 Engraving[1]

Just before a favorable decision by the Massachusetts Legislature to allow separation, the Eastern Argus newspaper in Portland, in arguing for the independence of Maine, asked the question: "Why should we be at the expense of legislating in another territory distant from one to three hundred miles? Is this an expense we do not pay? Then are we burdensome to old Massachusetts, and ought,

1 November 19, 1853 issue of *Gleason's Pictorial Drawing Room Companion.*

in honor, to relieve her from the tax. But if we really have to pay this expense ourselves, which I apprehend will be found to be the case, why might we not save all this extra expense of time and travel by legislating for ourselves at Portland, Augusta or Belfast? It is not the mere loss of time and money we experience by legislating in Boston; our money and our business center there, and so long as we are willing to look to another state, for a capital of trade, we never shall have one of our own. Instead of building up a town of our own, we send all our money and imposts out of the district; and we are so wedded to old habits that we first send our West India produce to Boston to pay tribute to the merchants there, before it can be bought by the retailers of Maine. This is strange; it however proves that, 'there is something out of joint.'"[2]

Having been designated an independent state in 1820, the Maine Legislature's session for 1822 brought up the issue of where to locate the seat of government.

Mr. Sprague, of Hallowell, argued in favor of a site in the Kennebec River Valley area, specifically, in Hallowell. He favored a location as close as possible to the center of population for the state. At that time, the population was concentrated in the region from York to Bangor.

> It was obvious from every consideration, that the seat of government ought to be near the center of population. Rulers were bound equally to relieve the wants and advance the welfare of every portion of the community. They should be placed in the midst of their fellow citizens, where the means of promoting the interest of all may most easily be learned. Its capital should enlighten and invigorate the whole; it must be the fountain of life to the body politic; and it was as preposterous to place it in a corner of the state, as it would be to find the human heart in the extremities.

Sprague went on:

> Was Portland then the place where the seat of government should be permanently established? That was substantially the question before the house. To show that it was not, it would be necessary to institute a comparison between that and other places. To be definite and intelligent, he would select one only, that with which he was best acquainted, and which he had the honor to represent.
>
> It was not desirable that the legislature should hold its sessions in a very wealthy and populous town. There would be an influence exercised unpropitious to fair and free legislation. He meant no disrespect to the citizens of Portland. He believed them as pure in this respect, as would be the other members, under similar circumstances from any section of the state, but it was incident to human nature. It was inevitable that they should use the means thus put into their hands to promote their own views and sub serve their own interest. Hence it was that almost every state in the union had

[2] Eastern Argus (Portland, Maine) April 20, 1819 p2.

withdrawn its legislature from the large and commercial cities. Massachusetts was an exception. While connected with Maine, it would have been difficult to have selected any other than the present seat of government, but such had been our experience of the evils of legislating at Boston, that we should in most respects, hardly be inclined to follow her example.

He knew it was objected, and that Portland was the only place large enough to afford suitable accommodations. Indeed, to hear some gentlemen, one would suppose that all other towns consisted of only some half a dozen houses thrown together by accident, and that the ingress of two or three hundred persons would be absolutely overwhelming. To speak for one only, that which he represented, he could assure gentlemen that their fears were wholly unfounded. It comprised a population of about 20,000 persons in a compact village, who with perfect ease in themselves would receive the legislature and all its attendants, and furnish them with every convenience. Of its resources, anyone might form an opinion by looking at its local situation and the condition of the adjacent country. It was the principle, the sole market of consequence, for a population of more of than 70,000 persons. The county of Kennebec was known and acknowledge to be the most highly agricultural of any in the state. In the exuberance of its soil and the profusion and excellence of its productions, it yielded to none in New England. The county of Somerset too, possessing a soil of great fertility, had advanced with rapid strides and was hardly inferior to Kennebec. Both of these counties, with a part of Lincoln, and a part of Oxford, poured a very great proportion of their surplus produce into the town of Hallowell. If gentleman would consider the variety and abundance thus produced, sufficient indeed to subsist thousands instead of hundreds, they would dismiss their apprehensions of creating a famine by a removal of the legislature.

It was sometimes said that they ought to continue the seat of government at Portland, in order to build up a great town. Tax their constituents three or more thousand a year to build up a great town! Were the legislators then sitting at some distant place, what would be thought of a proposition to take such a sum, for such a purpose from the territory?

He had often been told that the citizens of Portland had been liberal in their civility as to the members of the legislature, and in their exertions to accommodate the public, and it would be ungrateful to think of removing the seat of government. There was a direct appeal to these feelings. But were they at liberty to commit them in opposition to the interest of the state? Should they take thousands of dollars each year, from the public to gratify their inclinations, or partialities?

Representatives from Portland and Hallowell argued about which location would be the most cost effective considering travel costs and postage expenses.

Mr. Adams, of Portland, argued as follows:

> The gentleman from Hallowell meets us in the first place on the ground of expenses; and has gone into very ingenious and minute calculations, from which he thinks he has made it appear that the annual public expenses are two or three thousand dollars greater by the seat of government being in Portland, than it would be if it were in Hallowell. But he believes those calculations to be erroneous; it was an ex-parte statement and gave only a one-sided view of the subject.
>
> A large item of the savings in that gentleman's calculations was from the postage of letters and documents, which pass through the post office department. But when we consider that the post office in this town is the main distributing office in the state, and that a large proportion of communications passing to Hallowell, must necessarily go by the way of this post office; he believes that the savings on this subject would be much greater from having the seat of government in Portland, than it would be were it located in Hallowell. Another item of saving, which the gentlemen had brought forward, was in the expenses of living; he had estimated the price of board at $1.00 per week or less in Hallowell than in Portland. But he believes that facts on the subject are a much better guide than theory; and he could assure the members of the legislature, if they should find it less expensive living in Hallowell than in Portland, they would be much more fortunate than he had been when he had visited that town.
>
> There was another subject which ought to be taken into consideration in the calculation of expenses. It was admitted on all hands, that the legislature was very well accommodated at present with buildings to hold their sessions; and was without any expense at all to the state. But should the seat of government be removed, it would be necessary to erect public buildings. And for the purpose we should probably have to contract a debt of $50,000 or $60,000; the interest on the sum, provided we could obtain it at 5%, would amount to $3,000. Therefore, instead of our expense being diminished by removing to Hallowell, he believed it very rational to presume that they would be increased.
>
> Another argument, which the gentlemen had urged with great force of eloquence and ingenuity, was that the seat of government ought not to be fixed in a town situated on the seaboard because it was in danger of an invasion from a foreign enemy. And to give force to his argument the gentlemen had referenced the destruction of the capital at Washington during the last war. But the cases were totally different; the state could not be in danger of losing

> its public buildings for it had none to lose. And should property be destroyed in this town by a foreign enemy, the loss would not be upon the state but upon individuals in the town of Portland.
>
> Another reason, which had been urged in favor of removing the seat of government from Portland, was that there existed an 'outdoor influence,' which had an undue weight upon the proceedings of the legislature. He believed this argument was entirely without foundations. Was there a single member of the legislature who would not spurn at the idea of being controlled in the proceedings by interested individuals? This was an old story, it had been reported many times. It used to be the cry, when we belonged to the State of Massachusetts, that the capital citizens of Boston had an undue influence over the legislature; but as far as his experience extended on the subject, he believed that whatever individuals in the town of Boston had attempted to interfere in the proceedings of the legislature, it had always operated directly against the town; and should any individual in this town attempt to control the deliberations of the legislature, it would undoubtedly be attended with the same effect.[3]

A report was requested from a committee established to determine which site would best fit the needs of the Legislature during the 1822 session. On February 1, 1823, they reported "If it was expedient to locate the seat of Government near the seaboard, Wiscasset would be the most suitable place, but should it be thought advisable to locate it in the interior, they should give the preference to Augusta." Finally, they said: "That the seat of government be permanently located at Augusta and that the Legislature should meet there on the first Wednesday of January in the year 1827."

That should have been the end of it, but in the session of 1824, the Portland and Wiscasset areas representatives were not so easily persuaded.

Mr. Parsons of Edgecomb (near Wiscasset), thought that the deed for some thirty acres of land in Augusta, which was offered by some clever and patriotic citizens from Augusta, should not be accepted. He felt that the people of Maine should first be consulted. Mr. Adams, of Portland, agreed with Parsons and suggested that the question be postponed until 1827, the date previously suggested for the first session to be held in Augusta.

Mr. Williams of Augusta felt that since the committee had been duly appointed by the Governor and the report had been accepted by the Representatives, the Legislature was duty bound to accept the deed.

Adams responded that at the convention, which framed the Constitution, it was recommended that the Legislature meet in Portland and that if it were not for the agitation from the Kennebec delegation, this would not have been an issue. He then wanted to know who had asked for this deed of land in Augusta. His final comment was that the next request would be for an appropriation to put buildings on the Augusta

3 American advocate (Hallowell, Maine) February 16, 1822 p2.

land and that would seal the coffin. Adams felt that; "The instant we accept this deed, we pave the way to a call upon the treasury, and I am not willing to give my vote for such appropriations without the consent of the people." He then suggested again that the question be postponed until the 1827 session.[4]

The Senate disagreed with the House in their resolution to postpone the subject of Government location until 1827, and reasonably suggested that no pledge be made to move the government to Augusta until 1830, unless the finances of the state should be in a situation to authorize an appropriation to be made. The House re-considered and agreed with the Senate.

By 1825, the Senate did not choose to accept the deed, feeling that it was still too early to make such an important commitment.

In 1826, the Legislature questioned the advisability of holding the next year's session in Augusta as the original report of 1823 had suggested. Another committee had again suggested that the session of 1827 should be held in Augusta. Mr. Adams, of Portland, offered an amendment that struck out the word Augusta and replaced it with Portland. He had stirred a hornet's nest. Again, Adams prevailed and by a vote of 68 to 54, the name exchange had been approved.

Meeting once more in Portland in 1827, the issue finally came to a head. Three resolutions were presented. First, that on January 1, 1830, the seat of government should be fixed at Augusta in Kennebec County. Second, that the deed of Joshua Gage and others providing land in Augusta for use of the state should be accepted. Third, that the Governor should appoint a committee of three to look into the requirements for building on that site.

Adams, of Portland, again tried to wreck the plan. This time he asked that the date for fixing the seat of government be 1840 instead of 1830.

His effort was finally overcome and a date of January 1, 1832 was agreed on as the point in time at which Augusta would become the permanent seat of government for Maine. He did get a token in the act which stated the Legislature would meet in Portland until 1832.[5]

"Weston Hill" in Augusta was the site for the construction of the new state capitol. On February 24, 1827, governor Enoch Lincoln signed a bill establishing Augusta as the official seat of government for the state of Maine.[6]

In an interesting twist, the business men of Portland made a last-ditch effort to lure the Government of Maine back to Portland. In 1835, the Exchange Corporation started a building designed by Mr. Bond of Boston, which was to surpass all rivals in the New England area. The structure was made from Kennebunk granite, sporting a roof covered entirely with copper. One could almost mistake it for a State Capitol! The building was completed in 1840, but the State Government stayed in Augusta. It was used for Federal offices and courts over the years. The Exchange was gutted by fire in 1854.

4 Hallowell Gazette (Hallowell, Maine) January 28 and February4, 1824 pp2 and 3.

5 Eastern Argus (Portland, Maine) February 2, 1827 p2.

6 Maine State Legislature, *A Brief History of the Maine State House.*

Portland Exchange Building 1845[7]

Travel from Portland to the Kennebec Valley by stagecoach had begun long before Augusta had been selected for the Capitol site. As early as 1792, Caleb Graffam rode the post and conveyed newspapers from Portland to Hallowell. By December, of 1793, he carried passengers from Portland to Hallowell in a sleigh. This was just once every two weeks, but it made a good start![8]

In Part Two, we will cover the roads, stage lines, and taverns used to convey and accommodate travelers from Portland to Augusta.

7 Collections of Maine Historical Society Maine Memory Network Item 22425.

8 William Willis, *The History of Portland.*

Chapter Six

The Roads from Portland to Augusta

Although an early attempt was made to create a "Maine Turnpike", extending from Kittery to Augusta, most of Maine's roads were constructed in a piecemeal, as needed fashion. Unlike Massachusetts and New Hampshire, Maine had very few privately funded toll roads, and not all of those were between major locations.

Frederick J. Wood tells us the following in his book, *The Turnpikes of New England*, "Upon the severance of Maine from Massachusetts in 1820, the turnpike inheritance consisted of five roads: The First Cumberland Turnpike in Scarborough; the Bath Bridge and Turnpike, from Brunswick to Bath; the Wiscasset and Woolwich Turnpike; the Wiscasset and Augusta Turnpike; and the Camden Turnpike."

"Evidently, anticipating considerable business in forming corporations for that purpose, the new state adopted a comprehensive code of general laws closely following the Massachusetts practice, in which the procedure in such cases was fully outlined. The only radical departure from Massachusetts custom was in delegating to the county commissioners instead of to a committee appointed by the Court the duty of laying out the line of the road. But the labor of preparing these laws was wasted, for only one company was formed by the Maine legislature and that company failed to carry out its plans."[1]

That one company was the Maine Turnpike Association, which had already eliminated the planned section of road from Portland to Augusta by March of 1805.

The five Maine turnpikes which did carry through were all chartered in Massachusetts before 1820. This compares strikingly with the more than sixty turnpikes chartered in Massachusetts during the period 1796 to 1836. Those of particular interest to the eventual stage lines running from Portland to Augusta were on the north end, from Brunswick to Wiscasset and then Augusta.

The first section was chartered on February 22, 1803, and was to run from the courthouse in Wiscasset to the new Kennebec toll bridge in Augusta. This road was not even laid out until 1807.

[1] Frederick J. Wood, *The Turnpikes of New England*.

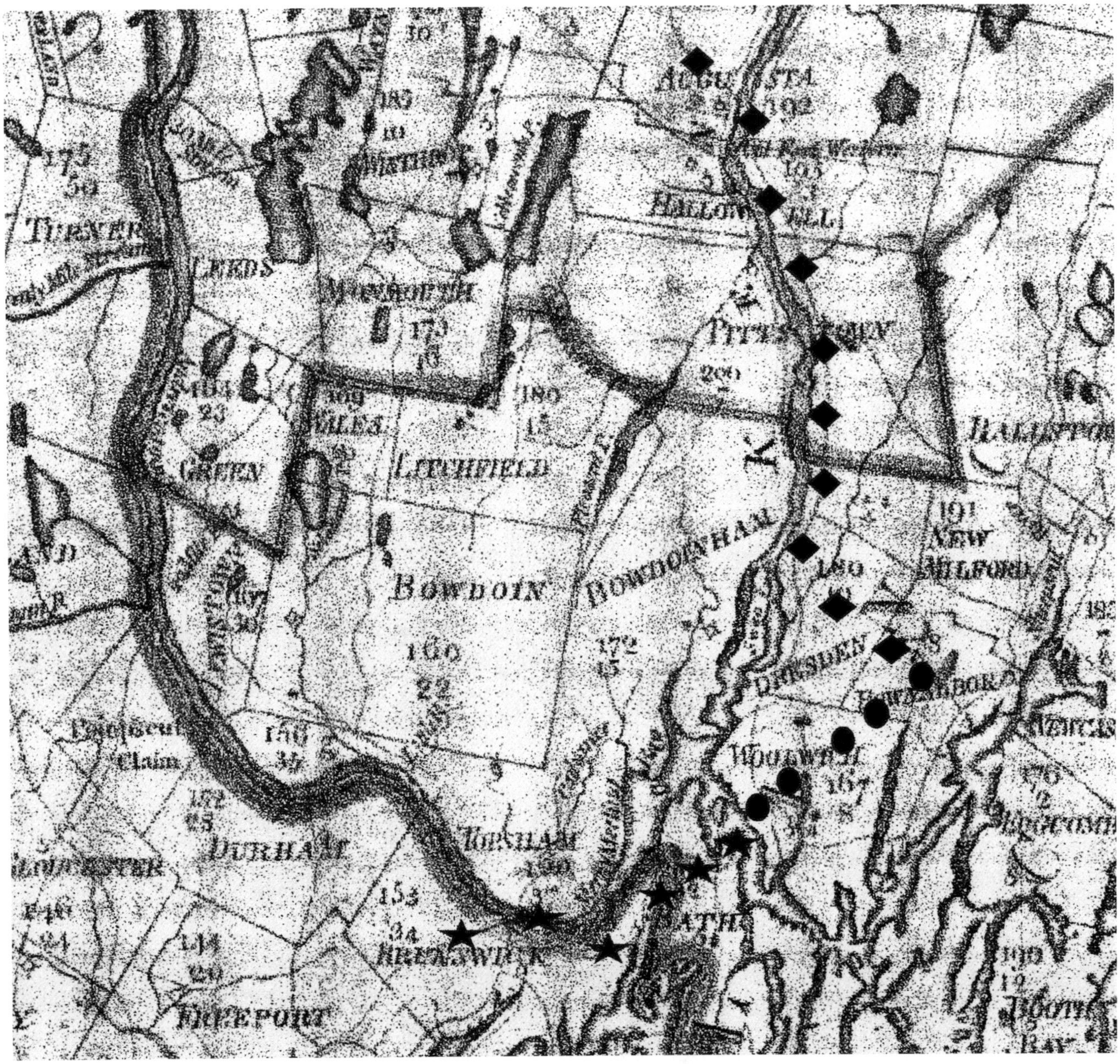

1802 Carleton Map of Maine Showing Turnpikes from Brunswick to Augusta
★ = Governor King's Turnpike; ● = Wiscasset to Woolwich Turnpike; ◆ = Wiscasset to Augusta Turnpike

The second piece was chartered on June 23, 1803, from Wiscasset to Woolwich which is next to Bath (across the Kennebec River.) It actually went through Woolwich to Day's Ferry, which crossed the Kennebec between Woolwich and Bath. The ferry had been operated by Samuel Harnden since before 1762. The Days took over in 1788 at which time the operation took on the Days' name. In 1837, a steam driven ferry was used and many stagecoaches passed over the Kennebec at that point. It is ironic that the steam engine, which was helpful to the stage lines at that time, would later put them out of business with the advent of the steam driven railroads.

Tukey's Bridge 1845[2]

The third piece of the puzzle was Governor King's Turnpike, from Bowdoin College in Brunswick to Bath, including a bridge over the New Meadows River. It was about eight miles long and was principally owned by Governor William King. The cost, including the bridge, averaged out at about $2,000 per mile. Much of the total cost was a result of the bridge construction.[3]

Now, one could travel from Brunswick to Augusta on a relatively smooth, hard road, except for the short hop over the Kennebec at Day's Ferry. From Wiscasset, the road followed the east side of the Kennebec River from Dresden to Augusta. The complete length for all three was 45 miles and these turnpikes brought Augusta closer to Portland. Of course, one still had to get from Portland to Brunswick to take advantage of the turnpikes. This portion of the road was also part of the King's Highway, as was the portion from Brunswick to Wiscasset.

[2] Collections of Maine Historical Society Maine Memory Network Item 22431, Charles Quincy Goodhue Drawing.

[3] Frederick J. Wood, *The Turnpikes of New England.*

The road to Brunswick started in Portland, and went through Falmouth, North Yarmouth, and Freeport. Getting from Portland to Falmouth was probably the most difficult portion as it involved either crossing the mouth of the Presumpscot at Martin's Point, or traveling southwest over Bramhall's Hill through Stroudwater before turning northeast to reach Falmouth. The latter must have been the only route for a long time.

Another possibility was to build a bridge across Back Cove between Seacomb's and Sandy Points. in 1791 a group of citizens petitioned the court to assist the town of Portland, by lottery or otherwise, to build a dam across Back Cove and erect grist mills on it. The government offered no help, so the citizens in the area incorporated to build a bridge at their own expense. By 1794, they had the charter and with diligent work, by September of 1796, a usable bridge existed. Stephen Waite was clerk for the Proprietors of Back Cove Bridge.[4]

[4] William Willis, *The History of Portland.*

The Second Martin's Point Bridge[5]

The Back-Cove Bridge was in due course named Tukey's Bridge, after its long-time toll gatherer, Lemuel Tukey. He ran a small tavern at the end of the bridge and did not appreciate the idea of ending the tolls as he had been instructed to do, so he continued collecting them long after he should have stopped. Legal action was required to finally have the tolls cease, and that was accomplished in 1837.

The next hurdle was the Presumpscot River in Falmouth. Again, citizens of Falmouth petitioned the court in 1758 for money to build a bridge at Smelt Hill, for crossing the Presumpscot. This time the court gave them the right to hold a lottery to raise the 1200 pounds required for its construction. The bridge was started in 1759 and finished in 1762, as a covered bridge operating as a toll bridge until 1789 when Cumberland County bought the bridge for public use at no charge. In 1801, a new bridge was built just downstream but several feet higher than the old bridge. Twenty years later it was turned over to the town of Falmouth.[6]

Another bridge across the Presumpscot, closer to its mouth, was completed in 1821. This was the Staple's Point Bridge which was constructed with a center draw and built by Seward Merrill. In 1837, the bridge was replaced with a covered bridge without a draw. In 1893, sparks from the Grand Trunk Railroad, running parallel to the bridge, set it on fire. Once again it was rebuilt but this time of steel and it was considered the finest bridge in Cumberland County.[7]

There was hope for an even more direct route as in 1807, the Massachusetts Legislature approved an "Act for Building a Bridge over the Presumpscot River, Near the Mouth Thereof."[8] This was to be constructed from Martin's Point in Portland to Mackey's Point in Falmouth.

5 Post Card Dated 1908.

6 Charlotte and Donald Wallace, *A Story of Falmouth, Maine.*

7 Ibid.

8 Gazette (Portland, Maine) March 9, 1807 p2.

Between the Embargo Act and the War of 1812, nothing further happened. It was not until 1828 that the bridge was finally built. It was a toll bridge and the fee to cross with a stagecoach was fourteen cents, the most expensive toll. In 1866, the bridge was destroyed by huge cakes of ice coming down the river.[9]

The final obstacle occurred in North Yarmouth, at the crossing of the Royal River. It is not certain when a bridge was completed to affect that crossing but by 1811, the bridge at the Lower Falls was in need of repair and widening. Public notice was given that it would be closed for three weeks but an alternate route over Gooch's bridge, one mile distant, was available. The signatory, Ammi Mitchell, now a Selectman of North Yarmouth, had the same name as the person who had petitioned the Massachusetts Legislature to create the Martin's Point Bridge some four years earlier.

The bridge at the Falls can be seen in the following photograph taken in the latter half of the 1800's. The rows of elm trees which so fascinated the travelers on the stages can be seen in the top center of the view. Justifiably, they felt this to be one of the most scenic spots on the route.

In 1822, another stage route to Augusta was opened from Brunswick. The trick here was to get across the Androscoggin River between

Bridge Across the Royal River in North Yarmouth[10]

[9] Portland Press Herald, Martin's Point Bridge, Charlotte Wallace.

[10] Collection of the Yarmouth (Maine) Historical Society.

1802 Carleton Map Showing Portland to Brunswick Route

Brunswick and Topsham. Ferries had satisfied this need from an early date, but were inefficient for stage transporting due to the time lost in crossing the river. It could take a couple of hours to load the barge, cross and then unload, a long time to travel less than a mile.

Mustard's Ferry had been in existence under some name from 1768. At that time, there was a road from Cathance to the Androscoggin at Mustard's, and from the New Meadows River to a point nearly opposite Mustard's. James Mustard of Topsham was licensed by the Court of General Sessions for Lincoln County to keep this ferry in 1784, but there must have been one kept at the spot long before. In the 1800's, the ferry landing was still discernible. In 1795 an attempt was made to have a bridge built across the Androscoggin River at nearly the same point by the towns of Brunswick and Topsham, and the town of Brunswick 'voted very generally, to build the one-half bridge across the Androscoggin

Androscoggin Bridge 1828 in Brunswick[11]

River, to begin near Dr Nuys mill.' A committee was chosen 'to see the matter carried on,' and the town also voted to raise 300 pounds for building the bridge, but not to access the money until leave to build it had been granted by the General Court. Nothing came of this attempt to construct a free bridge, but the next year certain persons were empowered by the Legislature to build a toll bridge from Nuys Mill in Brunswick to the Middle Rock and from thence to the rock below the 'Great Mill' in Topsham.

The act incorporating the proprietors of the 'Androscoggin Bridge' was passed and approved February 26, 1796. The incorporators were William King, Benjamin Jones Porter, John Dunlap, William Stanwood, Cutting Noyes, Amos Lunt, James Stone, John Merrill, Jr., James Wilson, Daniel Clark, Joseph Langdon, Ebenezer Emerson, Isaac Johnson, John Blanchard, John Merrill, Pelatiah Haley, Actor Patten, Benjamin Hasey, William Owen, and Theo Sims. A schedule of rates of toll was prescribed by the act, 'which shall be in force for thirty years, after that, subject to legislation.'

In March, an additional act was passed, fixing the number of shares at 500, at $8.00 each, and providing that no one person should purchase more than six shares within six days from the opening of the books, thus enabling persons of limited means to become shareholders and preventing control of the bridge being monopolize by a few individuals. The bridge was built during the summer of 1796, and a toll house was built in the fall. In the spring of 1811, the bridge was carried away by a freshet, and was rebuilt the same year at a cost of $5,591.42.

11 George Augustus and Henry Warren Wheeler, *History of Brunswick, Topsham and Harpswell, Maine.*

> On June 22, 1814, the toll house and greater part of the bridge were carried away by another freshet. They were at once rebuilt at an expense of $3,500. In the spring of 1827, the bridge was once again carried away and the directors voted to rebuild and to make the abutments of stone. Previously, they had been made of wood. The expense, including a toll house, was about $6000. Since 1827, the bridge has not been disturbed by freshets.
>
> In 1842, the bridge was destroyed by fire and rebuilt the same year. For some years previous to 1842, the bridge was a covered one. After that it was open. This bridge was repaired and made free on April 10, 1874. The value set by the appraisers was $2575. Eventually, it was owned by the towns of Brunswick and Topsham.

A second bridge was built in 1835, and it carried the name Bay Bridge as it crossed the Androscoggin where the Androscoggin flows into Merrymeeting Bay. Frank Connors in his article "The Bridge Over Merrymeeting", carried in the February edition of the 1977 Bowdoinham Daily Advertiser, gives its history:

> The Merrymeeting Bay Bridge was constructed in 1835, a crudely engineered, wooden span over the Androscoggin River between Topsham and East Brunswick. For more than 60 years, the half mile long bridge served as a shortcut between upper and lower Sagadahoc County, as a link between the agricultural upper towns and the expanding city of Bath.

Martin Hall of Bowdoinham, was the original contractor on the bridge. Owen's *History of Bath* says 'the bridge was constructed at a cost of about $20,000. Bath assisted in the venture and became a principle corporate partner in the enterprise,' by purchasing 160 of the 425 shares of common stock offered in the initial sale.

William Purington of Bowdoinham and William Randall of Topsham were among the first private stockholders. The list of proprietors included William King, George F. Patten, Joseph Sewall, David C. Magoun, William Richardson, John Patten, Daniel Marston, Charles Clapp, and Oliver Moses, all of Bath.

The bridge was 2700 feet in length, with wooden deck twenty-four feet wide. It was supported first on wooden pilings set twenty feet apart, and when the shifting river sands and ice and currents became a problem, rocks were added around the pilings to protect them. Still later, improvement projects on the bridge led to replacement of the wooden pilings with granite block supports.

Some of this old stone work is still visible from the Bay Bridge Road, off Topsham Foreside, and from the Bay Bridge Road on the Brunswick side. The bridge extended from Crooker's shore to Mustard's Island, then across the island and a short stretch of the Androscoggin to Topsham.

Unfortunately, the Bay Bridge was never a financial success. Maintenance

Bridge Over Merrymeeting Bay from Brunswick to Topsham[12]

cost always seemed to outstrip income at the toll booth. In 1837, when the bridge was barely more than a year old, Bath was forced to borrow money to meet an assessment against its stock for major repairs. In 1842, Bath's representative to the bridge corporation returned a statement to the town showing that on January 1 of that year, just $34.27 remained in the treasury after the toll keeper salary was paid and each of the stockholders was credited with a $2.00 dividend. Tolls averaged about $1200 per year, but in 1843, when tolls amounted to $1,300, repairs to the bridge cost $1,240.

In 1846, Bath authorize a loan not to exceed $10,000, 'to meet the assessment that may be made on the town's shares to repair Meerymeeting Bridge.' That amount was needed to cover only Bath's share! It has been estimated that in the 50-year history of the bridge, more than one million dollars was spent on the construction, repairs, and maintenance of the span.

By 1864, the last of the private stockholders became fed up with their investment and abandoned all interest in the span to the City of Bath, which bore the considerable burden of financial responsibility until 1875.[13]

[12] Courtesy Pejepscot Historical Society, Photo taken from Water Street, 1875.

[13] Bowdoinham Advertiser, February 1977.

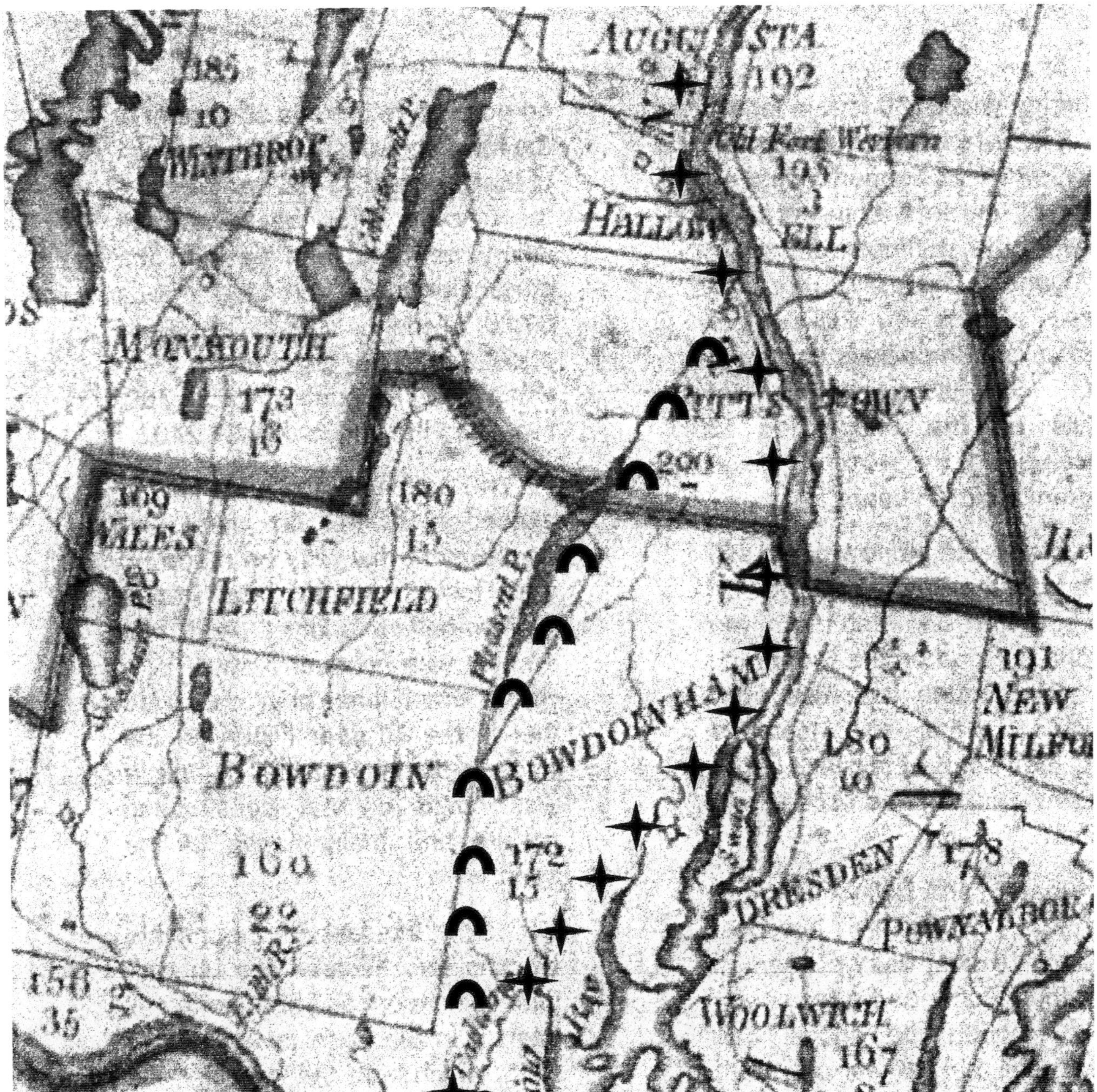

1802 Carleton Map Showing Brunswick to Augusta Routes

✦ = Earlier Route; ∩ = Later Route

The earlier route from Brunswick to Augusta seems to have crossed the Androscoggin at the Androscoggin Bridge, and proceeded through the Village at Bowdoinham, and then on to Gardiner tracking the west bank of the Kennebec River. With the advent of the Bay Bridge in 1835, the route followed close to today's highway 201, bypassing Bowdoinham and Richmond Villages, while tracking along the East shore of Pleasant Pond. The paths came back together at Gardiner.

The coastal routes as described above were later on in competition with another route to the Kennebec Valley. By 1826, a "New Stage Route," which we shall call the "Middle Route" was announced. The stage could now head north out of Falmouth, clip the northwest corner of North Yarborough, proceed to Pownal, pass through Durham, and cross the Androscoggin at Lisbon Falls. The route then angled over the corner of Sabbatus and proceeded on to Litchfield and Hallowell (not Augusta.) If this route sounds familiar, it is. This route is similar to State Route 9, from Falmouth to Sabbatus, and then pursues Rote 126 to Gardiner.

An alternative route would have been to turn off on what is now State Route 125 at Lisbon Falls, proceeding to Bowdoin Center, and then taking Litchfield Road to Dead River Road just below the Litchfield line. From there, they could turn north again on Academy Road, which becomes Hallowell-Litchfield Road, and continue on to Hallowell.

Building the Brunswick-Gardiner Road in Early Twentieth Century[14]

[14] Courtesy Litchfield Historical Society.

A Later Bridge at Lisbon Falls[15]

The article suggested that by examining Greenleaf's map of Maine, one could easily see how much shorter the new route would be by avoiding the great bend through Brunswick.[16] The even longer bend would have been on the original route which went to Bath and Wiscasset, then known as Pownalboro.

This new route needed two bridges, one for crossing the Presumpscot in Falmouth and another to cross the Androscoggin at Lisbon Falls. The bridge at Smelt Hill in Falmouth could also be used on this route.

The bridge at Lisbon, over the Androscoggin, was the result of an effort by local citizens of Lisbon. A meeting was ordered by the proprietors of the Lisbon Bridge in May of 1817. At that point, they were still trying to raise money for the project by selling shares of stock. Trouble stepped in quickly, as it was reported two years later: "We have to lament, that a great portion of the new bridge between Lisbon and Durham, and the new bridge between Lisbon and Brunswick, built at considerable expense, are now passing by us, while writing this article on the banks of Androscoggin Falls. The damage is great."[17]

By 1832, we find that the Androscoggin was flooding again. It had carried the Little River Bridge away and had seriously damaged the West Bend Bridge. There were by then two

[15] Courtesy Maine Historical Society Vintage Maine Images Item 5561.

[16] American Advocate (Hallowell, Maine) January 21, 1826.

[17] Bangor Weekly Register (Bangor, Maine) June 3, 1819. p3.

The Old Hallowell Road in Litchfield[18]

bridges across the Androscoggin at Lisbon. The transportation of logs cut in the forests of Maine posed a huge threat to bridges. The combination of large flotillas of logs heading down the Androscoggin to mills in the Brunswick area, and a sudden high flood of water would put any bridge at risk.

By 1828, yet another route had been opened. This one went from Freeport to Hallowell and Augusta. Since the old route was still active via Freeport, Brunswick, Bath, and Wiscassett, someone figured out that travel distance could be shortened by going directly from Freeport to Hallowell and Augusta.[19] This route became known as the Middle Road.

It was not clear until an ad, which appeared in 1831, defined the precise route as from Freeport to Durham to Litchfield, ending at Augusta. The ad was placed by a competing stage line.

This was not the final word on routes between Portland and Augusta. Much earlier, It was decided to use a road called the Back Route. This one went through Westbrook, Falmouth, Gray, New-Gloucester, Minot, Danville, Lewiston, Greene, Monmouth, Winthrop and ended in Hallowell.[20]

Fortunately, the bridge needed in Falmouth had been anticipated by Benjamin Winslow in 1787. Winslow was paid 180 pounds for his work. The bridge was later called Lambert's Bridge in

[18] Courtesy the Litchfield Historical Society.

[19] Eastern Argus (Portland, Maine) November 29, 1827.

[20] Eastern Argus (Portland, Maine) January 26, 1836.

honor of the Lambert Tavern which stood near it for many years.[21] This bridge led to the portion of Blackstrap Road which climbed its namesake hill.

Trying to picture four horses pulling a stagecoach up Blackstrap Hill makes one shudder! It is somewhat surprising that the real problem was coming down hills. In 1833, a serious stage accident was reported.

> When the Hallowell and Augusta stage, which goes on the upper route, via Green and Lewiston, was descending Blackstrap Hill, (Falmouth) between four and five o'clock yesterday afternoon, the horses became ungovernable, and Howe, the driver, was unable to hold them. There were eleven passengers, eight inside, two on the seat with the driver, and one, Mr. Brett from Dorchester, formerly from Turner, was on the top. A stage was ahead and, descending the same hill, the driver of which knew that Howe's horses were in the habit of running down the hill, and of course made room for him to go by. In going by the stage, to do which Howe was obliged to sheer out, he of course was thrown up on the side of the road. In making for the road again the bolt which fastens the fore wheels to the carriage gave way, the fore wheels and horses with the driver were detached from the carriage and the carriage was then precipitated forward, and thrown directly over, and down the hill, with great force. Brett was thrown off from the top of the stage and injured so severely that he lived only an hour. One of the passengers from Bridgewater, [Massachusetts] who was with the driver, had his thigh broken. Brett's wife was in the stage, who when looking out after the accident, saw her husband lying on the ground apparently lifeless. Her shrieks were then agonizing, notwithstanding she herself was somewhat injured. She struggled to get out of the stage, but ineffectually, and was at last taken out of the window. The mother of the passenger whose thigh was broken was in the stage, and was somewhat injured. All were more or less bruised, but none seriously. The trunks were broken open by the fall, and the clothing was spread in every direction. The driver escaped with but little injury. The horses, as we understand, ran about a mile, when the two pole horses became entangled, and fell and were dragged some distances by the leaders. One of the pole horses was much injured. The necessary remark arising from this accident is, if Howe's horses are in the habit of running downhill, and thus becoming ungovernable, no passenger who values his life, will trust himself to his care.
>
> Since the above was written, Mr. Howe informs us, that it is not a fact, that his horses are in the habit of running downhill, that they are under his management, and that in this particular instance the difficulty arose with one of the pole horses, who would not hold back. The accident did not occur on Blackstrap Hill but on the new road about five miles this side of Gray. We believe that Mr. Howe is a

[21] Charlotte and Donald Wallace, *A Story of Falmouth, Maine.*

> careful driver and gives general satisfaction.[22]

Two years earlier, the citizens of Cumberland County petitioned the Court of Sessions in Portland to find a better route from Portland to Gray than the one over Blackstrap Hill.

> The subscribers, inhabitants of the county of Cumberland, respectfully represent, that they, as well as the public generally, are subjected to great inconvenience, trouble and expense in their travel, by reason of having to pass over Blackstrap Hill, so-called, in the town of Falmouth, over which the county road from Portland to Gray and Windham now passes, and the same is greatly traveled; said hill being in some places very steep and rocky and more than one-half mile in length and extremely difficult of passage with carriages and loaded teams. They further represent that said hill may be avoided and a convenient road made by going to the eastward of said hill, commencing at or near Rand's Tavern, thence to Winslow's Tavern on the top of Lambert's hill. The distance being about 700 rods; and said contemplated route would avoid said Blackstrap Hill altogether and also shortening the distance about one-quarter mile. They also represent that said alteration in the road would be of great public convenience and could be made with little expense, (the proposed route being in every respect favorable for making a good road.) Your petitioners therefore prey your honors to cause a road to be located from near said Rand's tavern to said Winslow's tavern, on the same ground which was formerly located by Colonel Louis and others by order of this honorable court, and as is duty bound will ever pray. signed Elias Baker and 47 others.[23]

> In 1831 the county commissioners for Cumberland County ordered that the petitioners give due notice to all persons and corporations interested: that the county commissioners will meet at Thomas Winslow's in Falmouth on the second day of November next, at 11:00 AM when they will proceed to view the route set forth in the petition; and immediately after such view at some convenient place in the vicinity, will give a hearing to the party's and their witnesses; by causing attested copies of said petition and this order of notice thereon to be served upon the Town Clerk of Falmouth, and also by posting up copies of the same in three public places in said town, and publishing the same three weeks successfully in the Eastern Argus, the first of said publications, and each of the other notices, to be at least 30 days before the time of said meeting, that all persons may then and there be present and show cause, if any, why the prayer of said petition should not be granted. Attested: William T. Vaughn, Clerk.[24]

To this day there is still a dangerous combination of descent and curve on Blackstrap, which results in injury or death.

22 Portland Advertiser (Portland, Maine) October 8, 1833 p4.

23 Eastern Argus (Portland, Maine) October 25, 1831 p1.

24 Ibid.

The Old Wooden Toll Bridge between Lewiston and Auburn[25]

Back Route to Augusta

By 1797, the "Back Route," this time to Augusta, had been initiated. This route proceeded from Portland to Westbrook, to Falmouth, and then on to Gray, New Gloucester, Minot, Danville, Lewiston, Greene, Monmouth, Winthrop, and Hallowell. Since Portland, at the time, included only the neck, a corner of what was then Westbrook would be passed before entering Falmouth on what is now Lambert Road off of Auburn Street. The Presumpscot could be crossed using Lambert's Bridge in Falmouth, and then one would proceed north on Gray Road to the town of Gray. The next town encountered would be New Gloucester from where one could proceed to a stop at Danville, having clipped the corner if what was then Minot.[26]

The next challenge would be crossing the Androscoggin at Lewiston. Fortunately, the North Bridge at Lewiston had been built in 1823. Previously, only a ferry run by Paul Hildreth

[25] Courtesy of the Maine Historic Preservation Commission.

[26] Eastern Argus (Portland, Maine) January 26, 1836 p4.

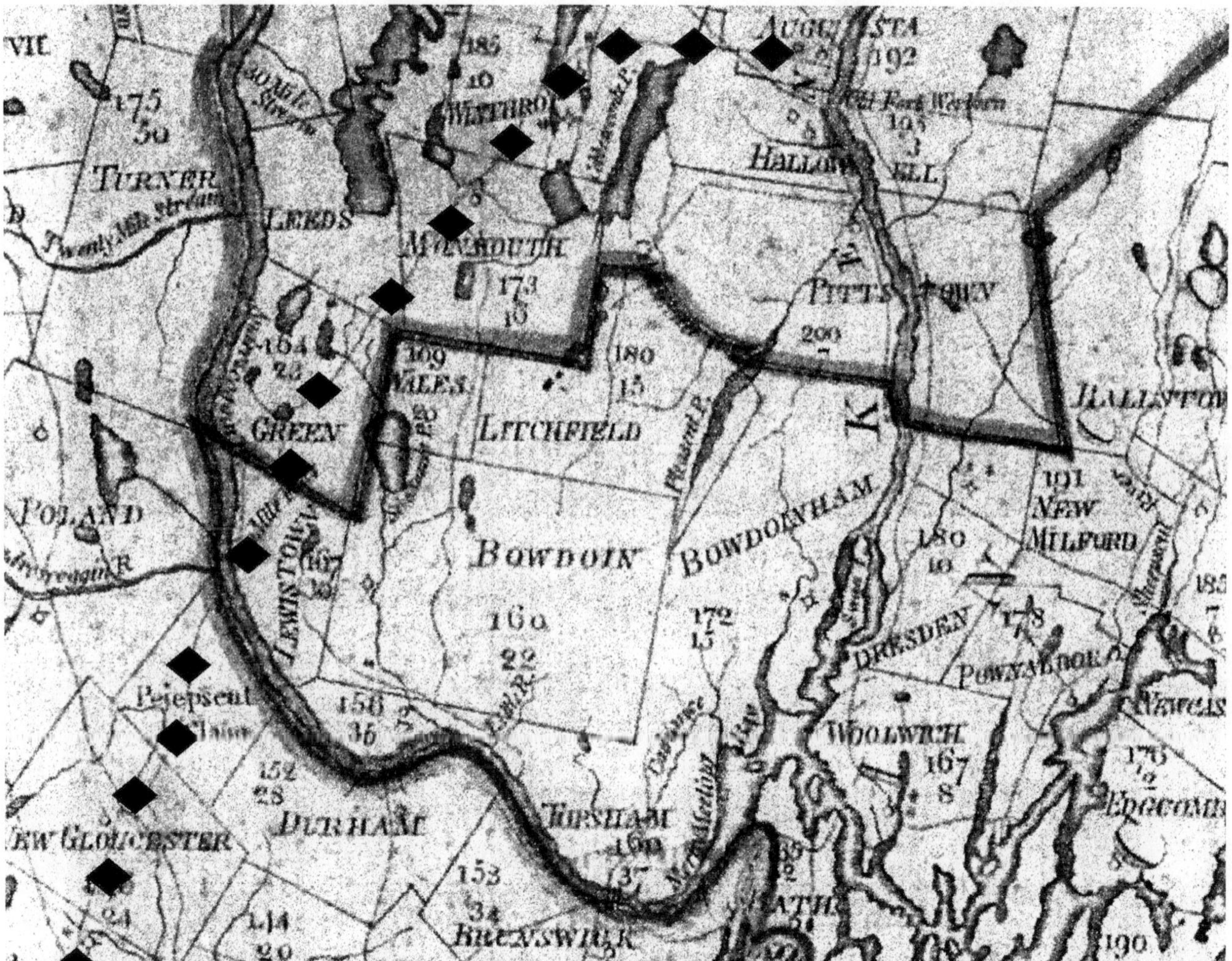

1802 Carleton Map Showing Route from Portland to Augusta via Lewiston

was available to cross. That method was begun in 1771 and lasted until a competitor moved in upstream in 1812. The Lewiston Bridge started out as a toll bridge, authorized by the Maine Legislature. The toll house was located on the now Auburn side of the river.

A flood in 1837 required a new bridge to be built, and this one included covered pedestrian walkways on both sides. Traveling by foot was a spooky proposition with one resident recalling "The darkness of those little covered sidewalks was no joke. It was almost as dark on those sidewalks by day as at night."

In 1865, it was finally decided to make it a free bridge, as it had become central to the life of the towns of Lewiston and Auburn.[27] Beyond the Androscoggin, there were no more rivers to cross.

It would be informative to digress for a moment to understand why Hallowell and Augusta, two cities of importance, within a stone's throw of each other on the Kennebec River, came into being. It is well known that Augusta and Hallowell were once joined as the city of Hallowell, named for Benjamin Hallowell, a Plymouth proprietor and merchant of Boston. The story of how this one town became two, as given by James W. North in his *History of Augusta*, is worth remembering.

The first hint of separation came with the attempt to establish two parishes within the town. It made sense as there were two villages involved. Fort was the designation given to the northern village containing old Fort Western while the southern village was called Hook representing the point at which the Kennebec slightly hooked to the west before proceeding south. Not being able to agree on the dividing line between the two parishes, it was suggested and agreed upon that three parishes would be better.

Having settled that issue, the town now tried to agree upon the best location for a bridge to cross the Kennebec River. Each village claimed that their preferred location was at the tidal head for the river That would be the furthest point which sea going vessels could achieve sailing up the Kennebec. A committee in the Massachusetts Legislature recommended that the bridge be built at Fort Western and a company was incorporated. The charter extended for seventy-five years from the date the bridge first opened for public use.

The two villages were represented by two newspapers, the Intelligencer printed at the Fort and the Tocsin printed at the Hook. As can happen, the competing newspapers fanned the fires of disagreement pushing the two villages farther apart. Separation of the two villages into two towns was desired by the Fort and rejected by the Hook. Following a petition from the Fort, the Massachusetts legislature approved the division in 1797, designating that Hallowell should keep its name and the new town should be called Harrington. Within a few months, having determined that the name Harrington was a little unwieldy, the name was changed to Augusta. The bridge was, of course, built at the fort.[28]

[27] Bangor Daily News (Bangor, Maine) October 30, 2016.

[28] James W. North, *The History of Augusta*.

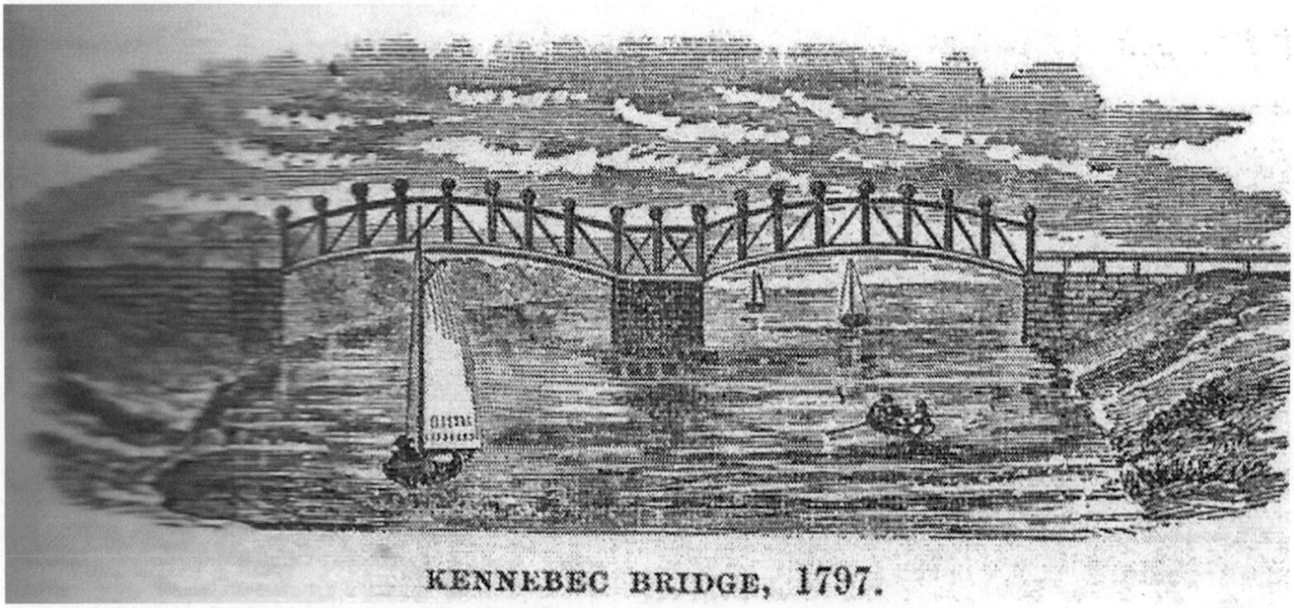

Kennebec Bridge at the Fort (Augusta) 1797[29]

Hallowell Chelsea Bridge[30]

[29] James W. North, *The History of Augusta*.

[30] Collections of the Maine Historical Society Maine Memory Network Item 29194.

Sixty-three years later, the citizens of Hallowell formed a corporation to build a bridge from that city to Chelsea, across the Kennebec. They wanted to compete with the bridge at Augusta and believed that it was a more direct route. The bridge saw very little use in its ten-year existence, and was washed out in the floods of 1869 and 1870. Try as it might, Hallowell simply could not keep pace with Augusta.

While all this construction of highways and bridges was taking place around the state, Portland was considering some improvements to its roads within the city.

> This is emphatically an age of improvement, and what with canals, railroads, steamboats, gas lights, etc., we shall ere long probably become the brightest as well as the happiest nation upon earth. We have not time to tell of the wonderful things that are doing abroad, but only to look at home and allude to the improvements that are making in our own town, which has just taken a start, as we say, as well as our neighbors. Some little zeal is now manifested, and some public spirit excited among our citizens, which we hope will neither slumber nor sleep, till whatever is necessary and useful for public improvements shall be fully accomplished—and first, of our streets. No town, we are confident, affords such facilities for MacAdaming or MacAdamizing the streets as our own—no town needs it so much, and in no place, can it be done and at so little expense. The experiment has been satisfactorily tried, at home and abroad, and answers the fullest expectations—so that in New York and Boston the city authorities are removing the pavements in several streets, for the purpose of adopting this novel and excellent plan of improving the street. We have just been conversing, too, with a gentleman from London, who says a complete destruction of the pavements is undergoing throughout that immense city for the sole purpose of MacAdaming the streets. We hope therefore to see the plan adopted here, and that one of our streets, at least may be completed the ensuing season.[31]

Encyclopedia Britannica tells us: "**Macadam**, is a form of pavement invented by John McAdam of Scotland in the 18th century. McAdam's road cross section was composed of a compacted subgrade of crushed granite or greenstone designed to support the load, covered by a surface of light stone to absorb wear and tear and shed water to the drainage ditches. In modern Macadam construction, crushed stone or gravel is placed on the compacted base course and bound together with asphalt cement or hot tar. A third layer to fill the interstices is then added and rolled. Cement-sand slurry is sometimes used as the binder."[32]

[31] Gazette (Portland, Maine) April 19, 1825 p2.

[32] Encyclopedia Britannica, britannica.com. Retrieved July 15, 2017, from Britannica.com website: <https://www.britannica.com/technology/macadam-road-construction>.

Chapter Seven

Building the Stage Lines Between Portland and Augusta

As early as 1793, we have it on good authority that Caleb Graffam Jr. began taking passengers by sleigh from Portland to Wiscasset, and even on to Hallowell.[1] Another item which suggests his involvement with staging is an article in 1798, stating that he was an inn holder in Portland. As previously noted, we frequently find stage operators running taverns or stage houses.[2] By 1800, he announced the opening of a new tavern at the sign of The Bunch of Grapes, near the Hay Market.[3] Caleb must have been the type of man who was very involved in his chosen city. In 1796, he was elected as a town officer for Portland.[4] His position was that of Constable, and in 1805, Deputy Sheriff.[5]

Caleb, Jr. was born in 1755, son of Captain Caleb Graffam, of Falmouth and Windham. Captain Graffam was a man of superior abilities, who took great interest in the welfare of Windham and was greatly respected by all who knew him. Apparently, his son followed in his footsteps.[6]

In 1810, Thomas Folsom of Portland took over the Columbian Tavern on Maine Street, formerly occupied by Caleb Graffam.[7] At age 43, Graffam married Mrs. Dyer in Portland and then died in Falmouth in 1813.[8]

At about the same time, in 1796, Joseph Stockbridge, a tavern owner in Bath, advertised for a new stage from Portland to Bath. His route took him from Portland to North Yarmouth, to Freeport, to Brunswick, and on to Bath. Leaving Portland on Sunday at five in the morning, passengers would not arrive in Bath until noon Monday. Another trip left Portland on Wednesday, ending up at Wiscasset on Thursday. Stockbridge ended his ad with a friendly P.S. "In the spring and the fall when the roads are bad, if they cannot be conveyed in a carriage, the subscriber will endeavor to furnish passengers with some other mode of conveyance." Sleighs became equally popular for group winter travel.

1 Eastern Herald (Portland, Maine) December 9, 1793 p3.
2 Oriental Trumpet (Portland, Maine) February 22, 1798 p1.
3 Eastern Herald (Portland, Maine) December 1, 1800 p1.
4 Ibid March 24, 1796 p3.
5 Gazette (Portland, Maine) April 8, 1805 p3.
6 Samuel Thomas Dole, *Windham in the Past.*
7 Gazette (Portland, Maine) January 1, 1810 p3.
8 Eastern Argus (Portland, Maine) November 20, 1811 p3.

The next stage operator who we find on the road from Portland to Augusta is Thomas S. Estabrook of Brunswick. He started in 1806 as a post carrier between Brunswick and Augusta, but was soon carrying passengers over that same route.[9] Just like Caleb Graffam and Joseph Stockbridge, Estabrook was also the owner of a tavern in Brunswick known as Estabrook's .We know that his tavern was in operation from at least 1815 to 1817, as he kept a ledger for his daily transactions during that time.[10] It is difficult to determine how long he provided passenger service by stage, as unlike many of his fellow stage owners, he did not advertise.

It is important to keep in mind that the early stagecoaches were not of the Concord coach variety. Usually, they would start as an open wagon, and if the route was successful, possibly morph into a stage wagon with some sort of cover to minimize the effects of the weather. The Concord coach did not come into being until 1827.

Colonel Thomas S. Estabrook was born in Lunenberg, Massachusetts in 1777. He also lived in New Ipswich and Hopkinton, New Hampshire, until 1801 when he arrived in Brunswick. Estabrook started a light infantry company, and by 1812, was a major of militia.

American Stage Wagon 1795[11]

[9] James W. North, *The History of Augusta.*

[10] *Estabrook's Tavern Day Book*, Bowdoin College Library.

[11] Isaac Weld, Junior, *Travels Through the States of North America.*

He was also a prominent fireman and for thirty years was college marshal. He was described as: "An honorable, upright, and genial man, fond of joke and story, and an excellent citizen."[12]

Estabrook died in 1859 and is buried at Pine Grove Cemetery in Brunswick. He had lived long enough to see the rise of stage travel, and its eventual demise under the wheels of the railroads.

Another Brunswick man, Henry McIntyre, is said to have been the first to drive a four-horse coach from Brunswick to Portland starting in 1803. He must have been born about 1782, as he was living in New Sharon in 1875, at age 93.[13] Many of the early stage men did not feel the need to advertise their stage routes as they were already well known from their being mail carriers or inn keepers. Neither was the case with McIntyre.

William B. Peters started a stagecoach line from Portland to Bath in early 1821.[14] Peters was first a cobbler on Exchange Street in Portland in 1817.[15] Next, he tried staging, but that seemed to last for only one year. His route was noted as "Start at Portland, through the towns of North Yarmouth, Freeport, Brunswick and then on to Bath." Peters ran his stages in conjunction with his livery stable, where he provided "Carriages, chaises and horses of the best kind." It is interesting to note that he did not associate himself with any of the taverns or hotels in town, but rather kept his stage books at his livery stable.[16]

At the convention for forming a constitution for the new State of Maine, Peters was appointed Sargent of Arms, reporting to the President, to assist the President in the preservation of order.[17] Later he took on the job of Deputy Sheriff under Sheriff Richard Hunnewell.[18] His decision to stick with a livery stable and forgo the stage business may have been a result of the new, more imposing stage company which was just coming into being.

Russel Stoddard, also of Brunswick, is the next stage owner to appear on the scene between Portland and Augusta. In December,1822, he and Samuel Hale, along with Israel Waterhouse, advertised the addition to their stage line from Portland to Augusta, of a stage leaving Portland every day at 6 AM. This stage would run from Portland through Falmouth, Cumberland, North Yarmouth, Freeport, Brunswick, Topsham, Bowdoinham, Gardiner, Hallowell, and arrive in Augusta the same evening.

Stoddard had been a tavern keeper in Brunswick while Hale and Waterhouse had built the Elm House in Portland. Together, they petitioned the Legislature of Maine to create a corporation to be known as the Maine Stage Company. The new company would provide stage service from Portland to Bath, and from Brunswick to Augusta. The Legislature passed the Act on February 10, 1823, and they were off and running.

[12] George Augusta Wheeler, *History of Brunswick, Topsham, and Harpswell.*

[13] Ibid.

[14] Gazette (Portland, Maine) January 9, 1821 p4.

[15] Ibid April 22, 1817 p3.

[16] Ibid January 9, 1821 p4.

[17] Jeremiah Perley, *The Debates, Resolutions, and Other Proceedings of the Convention of... .*

[18] *The Maine Register and United States Almanac for the Year of Our Lord, 1820.*

Stage and Four[19]

Stoddard ran a strange ad in March of that year entitled: *A Rare Chance for Speculators.*

> The subscriber being about to quit his present line of business offers his property in the Portland, Augusta and Bath Stage Line for sale, which consists of 27/90th part of the following, thus, six stagecoaches, three large double runner covered sleighs, one small covered sleigh, one small open sleigh, two single sleighs, one chase, one gig, with about 40 horses and proper harnesses, together with all hay, grain and other property which is on the lines.
>
> The conditions of sale are one-half purchase money to be paid on delivery of the property, the other one-half in three, six, and nine months, on interest and if the said property is not sold at private sale on or before the eighth day of March next it will then be sold at public auction at his house in Brunswick, at 11:00 PM to the highest bidder. For further particulars apply to the subscriber. Signed, Russel Stoddard.[20]

Was he already tired of the business or did he just want to see what value the business now had, six days after it had been formed as a corporation? The Maine Stage Company called their first meeting of the proprietors at the house of Russell Stoddard in Brunswick on October 7, 1823, so Stoddard must not have sold all his interest in the company by that time.[21]

An advertisement for an Accommodating Stage appeared in August 4, 1824, following the route from Portland to Bath. In that ad they mention that the driver had been driving on that route for five or six years. This could imply that the unincorporated prior company had been in

[19] *Business Directory of the Subscribers to the New Map of Maine*, 1858.

[20] Independent Statesman (Portland, Maine) March 6, 1823.

[21] Eastern Argus (Portland, Maine) September 9, 1823.

business since 1818/1819.The fare from Portland to Bath was just $1.50. Samuel Hale and Israel Waterhouse signed the ad as directors.[22]

Mr. John Robinson, a stage driver who had been in the employment of the Maine Stage Company, drowned by the upsetting of a boat in a squall just opposite McLellan and Turner's Wharf, in Bath, on May 6, 1825. He was only 23 years old, but he may have been the experienced driver mentioned in the previous ad, as stage drivers were frequently very young men.[23]

The meetings of the proprietors of the Maine Stage Corporation then moved to Portland and the next meeting, in June of 1825, took place at the house of Samuel Hale Esq. in Portland. L. S. Jackson of Brunswick was now the clerk of the corporation.[24]

In July of that same year, the Maine Stage Company advertised for a mail stage leaving Portland every morning at 4:00 AM for Augusta and Bath, arriving at Augusta at 3:30 PM. The fare from Portland to Bath for the faster mail stage was $2.00.[25]

Under the title *Something New*, it was announced in December of that year that Thomas S. Bowles has taken over the new and eloquent brick house on the corner of Federal and Temple streets, known as the Elm Tavern, as a well-furnished stage house where the stage books were kept, including those of the Maine Stage Company.[26]

[22] Independent Statesman (Portland, Maine) August 6, 1824.

[23] Maine Gazette (Bath, Maine) May 6, 1825 p3.

[24] Eastern Argus (Portland, Maine) June 28, 1825 p3.

[25] Ibid July 26, 1825 p3.

[26] Eastern Argus (Portland, Maine) December 20, 1825 p1.

A new mail route, from Portland to Hallowell, was announced by the Maine Stage Company in January of 1826. This new route would travel by way of Falmouth, Pownal, Durham, Lisbon, and Litchfield. Again, the two directors were listed as Samuel Hale and Israel Waterhouse. In Hallowell, the stage books were kept at the Dillingham's Hotel.[27]

By November of 1827, the company informed the public that they had commenced running a stage from Freeport to Hallowell and Augusta in conjunction with the stage from Portland. Apparently, Bath was not as important a destination as it had been before.

The company was doing well, and by 1829, had exceeded the amount of capitalization that had been allowed by the Legislature. The only answer was to petition the Legislature for an act to allow the Company to hold a larger amount of property. The company may have been anticipating an expansion of their stage routes as they were offering for sale a large stable and shed with a lot of land which had lately been improved by the Maine Stage Company. The interesting part was that this lot and improvements were located in Saco, well off the routes serviced by the company.[28]

An interesting article appeared in an Augusta newspaper in 1833. It was disclosed that: "None of the drivers in the employment of the Maine stage company, says the Portland Currier, drink any ardent spirits. There are nine in number on the lines from Portland to Augusta and from Portland to Bath. In consequence of some

[27] American Advocate (Hallowell, Maine) January 28, 1826 p1.

[28] Maine Palladium (Saco, Maine) March 25, 1829 p3.

inducements offered to them by the proprietors they entered into an agreement in May last for total abstinence for one year."[29]

In 1835, the company announced that they were "Building in North Yarmouth, on the new road, and near the Falls, to be ready to be occupied by the 1st of November next, a stage house," well calculated for a public house. This new house would be pleasantly situated nearly half way between Portland and Brunswick. It would be used for the mail stage as a stop for the passengers to breakfast when coming from Portland." The plan was to rent this house out to someone "well known or who could produce good recommendations for taking charge of such an establishment and keeping a good house." The stage business was going well and more investments were needed to both control the market and offer amenities to the travelling public.

Late in the year 1835, the Eastern Argus newspaper from Portland announced two important events. The first was the establishment of a General Stage Office in the city of Portland, where all the stage books were to be kept and where names of all passengers must be entered. This placed all the public houses in the city on an equal footing. In addition, a new stage line from Portland to Bangor, taking but one day to travel that distance, was added to the ever-expanding destinations for the Maine Stage Company. They used the familiar method of buying out the potential competition, and uniting the equipment purchased from Colonel Clapp, proprietor of the Cumberland House, to their own assets.[30] The Elm House was still being used for meetings of the directors and stockholders for the Maine Stage Company.[31]

Announcement of a stage over the Back-Route from Portland to Augusta appeared in an advertisement in January of 1836. The route went through Westbrook, Falmouth, Gray, New Gloucester, Minot, Danville, Lewiston, Green, Monmouth, Winthrop, and Hallowell before arriving at Augusta, all in the same day! This was not the Maine Stage Company, but rather a competitor who complained that the Maine Stage Company had added an opposition stage to the route, which had been held exclusively by the old proprietors of the Back-Route line. They even asserted that they would keep the one way fare at the old level of $2.50 as they felt that price was sufficient to sustain their enterprise.[32]

Today's stock holders would be shocked if they received a letter from a company indicating that they must pay an assessment in order to keep possession of their shares. L. T. Jackson, then treasurer of the Maine Stage Company, notified all shareholders that an assessment of eleven dollars and forty cents would be due on each share by February 12, 1837. Failure to make that payment would result in the delinquent shares being sold at auction.[33] By May of that year, notice was given to the public that the fare from Augusta to Portland would be reduced to $2.50, probably to be competitive with the Back-Route

29 Portland Advertiser (Portland, Maine) February 18, 1834 p2.

30 Eastern Argus (Portland, Maine) December 5, 1835 p2.

31 Portland Advertiser (Portland, Maine) January 19, 1836.

32 Eastern Argus (Portland, Maine) January 26,1836 p4.

33 Portland Advertiser (Portland, Maine) February 21, 1837 p3.

stage line.[34] A month later they announced that they had changed to the Mansion House in Augusta as their new stage house, where the stage books would be kept.[35]

Business went smoothly for the next few years and in January of 1846, the company was considering the prospect of extending a mail route to Montreal:

> **The Express for Sherbrooke and Montreal.**
> The express for Montreal alluded to in Saturday's paper was run from this city (Portland) to Norway, forty-three miles, by Mr. Hobbs attached to the Maine Stage Company, in two hours and forty-five minutes, gaining 15 minutes upon the time assigned in the estimate upon the way bill of the arrangements.
>
> Considering the condition of the roads, which are considerably drifted on some portions of the way, this may be considered a good indication of its success and leads us to expect it will reach Montreal within 26 hours, the time estimated by those conducting it, as necessary to run from the city. The next six miles were run in nineteen minutes. Mr. Waterhouse, of Paris, takes charge of it from Norway to Montreal.[36]

The amazing point of this announcement was that the sleds had to be moving in excess of 15 miles per hour in order to achieve the listed times! Of course, if the snow pack was good and the resulting path smooth, it does seem possible.

Walter Emerson, in his book *The Latchstring, Maine Woods and Waters*, gives a wonderful account of what he calls an "international sporting event." Hobbs, Bodge and Waterhouse were the three drivers who took on the challenge of taking the British Mail from Portland to Montreal faster than their counter parts could carry it from Boston to Montreal.

> Then, as now, the winter port of Canada was in the states, and Portland with her better and nearer harbor was beginning to loom large as a rival of Boston, in spite of the fact the latter was the larger city and more important commercial center. The British Government was concerned in getting the foreign mails to Montreal in the quickest possible time after they were landed by steamer on this side of the Atlantic. Boston, as was Boston's way in those days, said, ''Why, our way, of course. - up across New Hampshire and northern Vermont to the St. Lawrence.' Portland offered the Maine route, - up through Cumberland, western Androscoggin, and Oxford Counties, across the northern wedge of New Hampshire, then into the Province of Quebec to Montreal.
>
> They had a preliminary trial—a sort of scoring—and Boston won. But this was because the mails were delivered in Boston, where they had a long start, due to the wreck of the Eastern Railroad train which carried the bag to be sent by way of Portland. Then, the Maine dander was up, and the real

34 Ibid May 16, 1837 p3.

35 Age (Augusta, Maine) June 7, 1837 p4.

36 Portland Advertiser (Portland, Maine) January 27, 1846 p3.

race was on. Bear in mind that the transportation was by good. old-fashioned horse-power along the established stage routes, Portland contended that a fair test required at least an even start. So it was arranged. that a small packet should take the Maine bag from the English steamer somewhere out in the Gulf of Maine, southeast of Cape Elizabeth. Then the steamer was: to proceed to Boston. and the grand drive was to begin under conditions more nearly equal.

Great were the preparations down here in Maine. They went on for months. Morgan horses were obtained wherever possible and they were trained to the hour. There were relays from three to five miles, and in several instances, according to the hills, the run for one horse was only two miles. For weeks all along the route they were exercised and rubbed, and rubbed and exercised. No thoroughbred racer was ever more carefully prepared for the Derby. The expenses of these preparations, and the race itself, were paid by popular sub-scriptions along the road, all the way from Portland to Montreal. Everybody had native pride in the affair. All were interested, many excited.

Who could tell then when an ocean steamer would arrive? All that was known on this side was the date on which she was expected to sail. So the little packet tossed outside for days, and up in the Portland Observatory on Munjoy Hill sat a lookout day and night, ready to give a prearranged signal whenever the steamer herself, or the packet. coming In, should be sighted. Well down on Congress Street were a pair of horses and driver, ready to start out along the route to notify the real participants to hitch up and be ready, the mail had arrived. As a matter of fact, there were two or three relays of this notification committee, and they reached Paris Hill before the racers themselves caught up.

Selected to make this historic drive were the three best stage-drivers on the roads from Portland to Montreal-Hobbs, and Bodge, and Waterhouse. And in. this order, for each was chosen for the section which he knew best -Hobbs from Portland to Gorham, New Hampshire; Bodge from Gorham to St. Hyacinthe; and Waterhouse from St. Hyacinthe to Montreal.

Hobbs, and Bodge, and Waterhouse—personages in those days. Of course there were senators and governors, and the like, but they were mythical, read. about, never seen in the flesh. But stage-drivers—ah, they were real and important!

At last! The mail arrived. The signal was waved from the Observatory to the waiting outrider, who was soon away, and Hobbs, with his Morgan thoroughbred and skeleton sleigh, was in waiting at what are now known as the Grand Trunk docks. The mailbag was strapped to his back, so that in case of accident to the sleigh, he could jump on the horse and ride on without delay, or, should accident happen to both horse and sleigh, he could run on himself to the next relay. And away dashed the hardy Hobbs on his wild ride, probably amid the cheers of the crowd. Up India Street and out by Allen's Comer, where a fresh team awaited him. It took less than a minute to make the change, for along the road where the warnings had been given, the horses were all harnessed and ready for the run. There were half a dozen changes between Portland and Ricker Hill, Poland, one of which was at Brown's Tavern, at Gray Comer, famous among the early stage-road inns of Maine, kept—and tradition says well kept—by the father of the late J.B. Brown, who might himself be called the father of modern Portland. The buildings are still standing. Hobbs had the hardest part of the ride. A few days before the start there was a good sized snowstorm along his section, and the country roads were in poor condition, some being badly drifted. The race proceeded without any extraordinary incident until the foot of Shaker Hill, on the northern side, was reached. There the horse plunged through a big drift. But the sleigh stuck. The tugs and other parts of the harness snapped, and the racer was soon clear of the rigging. But not clear of the spry and resourceful Hobbe., for he, held to the reins, jumped the dasher, pulled in the steaming steed, mounted, and rode the long hill to the Ricker House at a smart canter. Here he was received with due ceremony by Sam Mills, and Wentworth Rickers, trusty horseman, who had a fresh team in readiness. It seems that Hobbs's mishaps were bunched in this short section of the ride, for Sewel! Brackett, who stabled and rubbed down his horse, says that on turning too sharply into the yard, the rider was thrown and pitched into a big drift. As before, he was uninjured, and little time was lost; Meantime, the precaution of strapping on the mail-bag had justified itself.

Thus the ride went on, up and down the hills of Oxford, with now and then a cheer at the taverns and stores, and, so far as I can learn, without untoward event. The going got much better beyond Bethel, and Hobbs crossed the State line and rode into Gorham, New Hampshire, man and horse in good form. Here Bodge took up the running, under much the same conditions, except that he had to wait for fresh teams to be harnessed, as there was no means of sending word ahead after the advance guard had been passed at Paris Hill. Waterhouse caught on at St. Hyacinthe, Quebec. He had the shortest of the three runs, and the best roads, and rode in record time into Montreal, where he was given the keys and freedom of the city and all that• goes with them, which—they tell me—means much or means little in Montreal , according to the capacity of the hero, Let us hope that he was a modest hero, and bore the responsible burdens of hospitality wisely and not too well. The newspapers of Canada made much of the event, and the arrival of Waterhouse was ceremoniously announced in the Canadian Parliament in session at the time.

But what of the Boston riders? Hadn't been heard of. Nobody seemed to

> know where they were—or cared. There's nothing quite so uninteresting in Montreal— or anywhere else—as a. loser. The Maine route and the Maine drivers had won, and won handsomely. They were the fellows! You will get some idea of the extent of the winning when you appreciate that at the moment when Waterhouse delivered his mail-bag at the Montreal Post -Office, the mail by way of Boston was somewhere between Newport, Vermont, and St. Johnsbury. The distance by the carriage road of those days from Portland to Montreal was about three hundred miles. The run, was made in twenty hours, or an average of about fifteen miles an hour.

By July of that year, things had changed dramatically. In the stockholders meeting for that month, they were going to consider selling the property belonging to the corporation.[37] Apparently, they stumbled along for another two years, so that in their meeting in May of 1848, they planned a special meeting to give instructions for the future management of the company's property.[38]

By August of 1848, two new directors, St. John Smith and H. J. Little were offering for sale the large stable on Chestnut Street recently occupied by the Maine Stage Company.

Two more years passed and in 1850, a meeting was called to advise the stockholders of the progress made in disposing of the company's property.[39] A final meeting was held at the Tontine House in Brunswick, on January 11, 1851, for a final report from the directors.[40]

The Maine Stage Company had closed their doors in 1849, with a book value of $15 per share versus a par value of $100. 600 shares were then outstanding owned by 50 different individuals. The stockholders were spread from Augusta and Durham, to North Yarmouth and Brunswick. Most lived in Portland but the company headquarters seemed to be in Brunswick.[41]

At that point the Maine Stage Company no longer had the guidance provided by its founders. Israel Waterhouse died suddenly in December 1839.[42] Samuel Hale died just six months later.[43] Stoddard lived until 1851, but had been out of that business for some time, even moving to Greene several years before his death.[44] Some have suggested that the diversely located stockholders simply could not keep adequate tabs on the company's day to day operations. While that could be a factor, the nail in the coffin was the arrival of the railroads to Maine.

A word about the character of these men who founded the Maine Stage Company is in order before moving on to the taverns. Waterhouse, Hale and Stoddard all appear in print at least once with an Esquire attached to their names. It is doubtful that they actually studied law, but in that day, the title was common for men of

37 Ibid July 28, 1846 p3.

38 Ibid May 16, 1848 p3.

39 Portland Daily Advertiser (Portland, Maine) August 32, 1850 p3.

40 Ibid January 10, 1851 p3.

41 J. Koppel, Editor, *Origins of Shareholder Advocacy.*

42 Gospel Banner (Augusta, Maine) December 7, 1839 p3.

43 Portland Advertiser (Portland, Maine) July 14, 1840 p3.

44 Maine Farmer (Augusta, Maine) June 29, 1853 and Pine Grove Cemetery, Brunswick, Maine, Tombstone.

prestige in their communities. Remembering how careful communities were about granting licenses for a public house, and realizing that all these men were wealthy merchants, it is not surprising to find their names so honored.

In 1822, Israel Waterhouse invited all of the Portland Light Infantry to dine at his tavern.[45] Once the Elm House was built, Waterhouse offered for lease his tavern on Middle Street.[46] By 1830 he was part of a committee to enquire into the possibility of making Back Cove Bridge in Portland a free bridge.[47] Of course, it would have been advantageous to his stage company to do so. In 1833, Waterhouse was an Overseer of the Poor,[48] and in the same year was part of an inspection team to check all buildings, to see that back yards and cellars were cleared of all filth, dirt, and rubbish. His area to inspect was Ward 4. Spring was cleanup time in Portland![49]

Waterhouse was also a staunch member of the Third Parish of the Congregational Church, and as such, was called upon to discipline members who had fallen out of grace with the congregation.[50] His obituary, dated December 7, 1839, includes the comment: "Mr. Waterhouse was an active business man, and an extensive mail contractor, and his loss will be felt in this respect very much."[51]

In 1824, Samuel Hale dissolved his business relationship with Mason Greenwood, probably to concentrate more fully on the Maine Stage Company.[52] In 1831, he petitioned the Legislature to incorporate a manufacturing company.[53] The latter was likely due to his getting out of the stage business as his name no longer appeared as a proprietor for that company. Notice of Hale's death appeared in July of 1840 where it was mentioned that "He was long known as an extensive mail contractor."[54]

Russel Stoddard's course through life was somewhat more difficult. In 1840, he apparently defaulted on a mortgage against a house in Brunswick, and upon his death in Greene in 1851, was declared financially insolvent. He did, however, live to age 84.

Another stage line running from Augusta to Portland, and even on to Dover, where it connected with stages to Boston, deserves mentioning. The Union Stage Line seems to have started in late 1831, and at that time was offering a travel time of two days from Augusta to Boston.[55]

An interesting story about one trip appeared in 1832: "It is stated to us by Mr. Dearborn, the driver of the Union Line of Stages, that about 6:00 last evening, he overtook between Kennebunk and Saco, two Irishman, one of whom marched up and caught hold of the reigns of the leaders, but said not a word. The driver laid his whip across the Irishman, who held on, until he, giving up the reins to one of the passengers on the box, jumped off and

45 Eastern Argus (Portland, Maine) June 11, 1822 p2.
46 Ibid May 4, 1824 p3.
47 Portland Advertiser (Portland, Maine) April 20, 1830 p2.
48 Eastern Argus (Portland, Maine) April 1, 1833.
49 Portland Advertiser (Portland, Maine) May 20, 1833 p3.
50 Christian Pilot (Portland, Maine) February 13, 1834 p3.
51 Gospel Banner (Portland, Maine) December 7, 1839 p3.
52 Independent Statesman (Portland, Maine) April 30, 1824 p3.
53 Portland Advertiser (Portland, Maine) March 15, 1831 p1.
54 Ibid July 14, 1840 p3.
55 Eastern Argus (Portland, Maine) November 22, 1831 p1.

knocked him down. The Irishman was soon up, and the driver knocked him down again, but in the meantime, he called for help from his companion, but called in vain, the other Irish man looking on without daring to take part. The driver then remounted his box and drove off. It is thought, they mistook this for a mail stage, or had some design on the passengers, but were frightened by the firm resistance of the driver."[56]

The Union Line seems to have lasted only a few years. Their final route to Augusta was a little unusual, passing through Freeport, Little River Village, Bowdoin, Litchfield, Litchfield Corner, West Gardiner, and Hallowell.[57] In their book, *Illustrated History of Kennebec County, Maine*, Editors Kingsbury and Devo state that the Union Line was started by Solomon Brown in 1830 and sold in 1848 to Crowell and Baker. A few years later, it had diminished to the point where it ran only from Augusta to Little River.[58]

Finally, there was the "Back Route" line from Portland to Augusta, started and run by Thomas Longley.

> Longley came to Greene from Waterford sometime between 1808 and 1810, and soon after coming to town. became widely known as the popular stage driver on the 'back route' between Portland and Augusta, via Gray, New Gloucester, Auburn [Danville,] Lewiston, Greene, Monmouth, and Winthrop: and was the first driver on this route, beginning in 1814, which he followed for 20 years. He had established himself on the place which later became the Phillips Rackley homestead now occupied by E. C. Rackley. Here he kept tavern which in those days became a jovial center around the generous fireplace. He had a large stall of horses. First the trip, made every other day, was with two horses: then with business increasing, the 'Concord coach and four' were used. Mails were also carried and the post office was then at Greene Corner in the old Clark house, built by Benjamin Parker; Luther Robbins Esq., was postmaster. Thomas Longley was a tall man of pleasing features, commanding in appearance, courteous and obliging, gentlemanly in manners, and remarkably well adapted to the business of staging; and for skill 'to curb the steed and guide the wheel' was hardly surpassed by Plato's 'Grecian Youth.'
>
> Thomas Longley was a great story teller; and it is said the aroma of his many narratives lingered along his route long years after he had surrendered the 'whip'. He had a tenacious memory; and while he never made a memorandum, was entrusted with many errands along the way, 'and never failed to make good', it was said.[59]

In 1824, he and Thomas Fuller notified the public that they had added more horses and an "elegant carriage" to their route. At that time, it was called the "Portland, Hallowell and Augusta Upper Mail Stage Route."[60] Following Thomas,

56 Christian Intelligencer and Eastern Chronicle (Gardiner, Maine) May 3, 1832 p3.

57 Zion's Advocate (Portland, Maine) January 23, 1833 p3.

58 Henry D. Kingsbury, *Illustrated History of Kennebec County.*

59 Walter Lindley Mower, *Sesquicentennial History of the Towne of Greene Androscoggin, Maine 1775 to 1900.*

60 Hallowell Gazette (Hallowell, Maine) May 19, 1824 p3.

his son Benjamin ran the line for several years, and he in turn was followed by Luther Kimball, Lewis Howe, and then Almon Bodge.

As an interesting side note: "Mrs. Thomas Longley, of Greene presented the Governor, yesterday with a cheese, big enough to satisfy the most inveterate lovers of that article. Its dimensions are in diameter, two feet five inches, in thickness eight inches, and it weighs 287 pounds."[61]

[61] Age (Augusta, Maine) March 10, 1840 p1.

Chapter Eight

Stage Stops and Taverns on The King's Highway to Wiscassett

Having covered the inns of Portland in Chapter Four, we can now move on to the first possible stops after leaving Portland heading North or East, as the locals would say.

Falmouth

There were three routes through Falmouth for the stage lines. The first was along the King's Highway crossing Back Cove Bridge, and then heading toward Smelt Bridge on the Presumpscot. This would take you past Bucknam Tavern on State Route 9 or Middle

West Falmouth-Starting Point for the New Gray Road with Blackstrap Mountain Tower in the Distance[1]

[1] W. W. Clayton, *History of Cumberland County, Maine.*

Bucknam Tavern[2]

Street. The second started off to the west, over Portland's land connection, then north past Lambert's Tavern, heading for Lambert Bridge across the Presumpscot, then to the top of Blackstrap Mountain and Winslow's Tavern. The third would split off at West Falmouth Corner and then proceed on the New Gray Road toward Gray, stopping at Hall's Tavern or Washburn Tavern. In this chapter, we will examine taverns along the first route, the King's Highway.

Built by Samuel Bucknam before 1793, this tavern served the community of Falmouth as well as the travelers along the King's Highway. Bucknam was licensed to run an inn for the years 1793 through 1796 and for 1798. Samuel died in 1814, having been bedridden for seven years. His third wife, Huldah, was licensed to run an inn from 1815 to at least 1820. Presumably, other family members were in charge in the interim. Samuel Bucknam may have moved to North Yarmouth for a time, as someone by that name was a party to a petition concerning 100 acres along the Royal River in 1804.[3] Buknam had been a Town Officer for Falmouth in 1785.[4]

2 Personal Collection of Marge Devine, Falmouth, Maine.

3 Gazette (Portland, Maine) October 1, 1804 p4.

4 Falmouth Gazette and Weekly Advertiser (Falmouth, Maine) April 9, 1785 p3.

Now, the Bucknam Tavern in Falmouth was only a little over six miles distant from the center of Portland, and would have been a little closer than a normal distance between stops of nine miles, so one would not expect the stages leaving Portland to stop here on a regular basis. Adverse conditions such as weather or road conditions might call for an earlier rest to change horses and to offer a short break to the passengers.

Today, the Bucknam Tavern is in very poor condition, possibly near collapse. It is owned by people from out of state who obviously have no interest in its survival or its adverse effect on the rest of the neighborhood. An article entitled "That's My Dump," by Chad Frisbie, appearing in 2011, stated: "The walls wouldn't tell me anything, but the neighbors had tales to spin. Legend has it that a guest at the Bucknam Inn once went insane and decapitated his companion. He chucked the guy's head out an attic window. Since then, whenever a new pane is installed in that window, it shatters soon afterwards – cursed by the flying cranium.

Now it seems the curse has spread to nearly every window in the house. White curtains blowing out of broken panes waved me off like ghosts. Whether caused by hocus pocus or pure neglect, the Bucknam Inn's deterioration may soon preclude any opportunity to salvage this unique example of Falmouth's heritage. Let's not throw the house out with the head!"[5]

[5] *The Bollard*, March 7, 2011.

North Yarmouth

If we were traveling on the King's Highway heading northeast toward Wiscasset, our next stopping place would be North Yarmouth, approximately eight miles away from the Bucknam Tavern in Falmouth. Here we would find a wide variety of taverns. The earliest principally served the locals, while those coming later were located to serve the traveling public as well. Some were around long enough to supply the needs of all the citizens who might come their way. To get the most business, a tavern or inn needed to be located on a stage route. Over the years, the stage roads through North Yarmouth kept improving by shifting the physical roads to make the routes shorter and less difficult for the stages to maneuver. It was probably easy to change the route from one street to another, but taverns were harder to move!

Early in the eighteenth century, we would have found the taverns of Captain James Parker and Jacob Mitchell. Parker's was the first, in use by 1722, but it probably was in place too early to be a stage station. Mitchell's was definitely a stage stop.[6]

James Parker was born in Groton, Massachusetts in 1687, and came to North Yarmouth by 1720. Parker died in 1732. Again, being a tavern owner required active involvement in the community, and Parker was no exception. He was involved with protecting the woods from random cutting by men who were not part of the community. He also helped

[6] Augustus Corliss, *Old Times*, Vols. 5-8.

Jeremiah Mitchell Tavern—North Yarmouth[7]

correct a poor job done on the laying out of the town lots and was on a committee to build the first meeting house.[8] No doubt, his tavern was the local meeting spot and probably housed the occasional traveler in the area.

Jacob Mitchell Tavern has a fascinating history. It started as a house or garrison within a stockade, built in 1726. Located less than 100 feet from the Royal River, it featured a secret tunnel to allow the occupants a way to escape to the river from an attack by fire or enemies.

Being on the first stage road through town, as the stagecoach business picked up, the stockade was torn down and the house enlarged to accommodate stage passengers. The original structure featured the usual clapboard construction while the new addition, on the north end, was clad with shingles. The Jacob Mitchell Tavern stood on Pleasant Street until 1923, when suffering from the effects of too many Maine winters, it was burned down.

Jeremiah, grandson of Jacob Mitchell, was born in North Yarmouth in 1790. He was a Sargent in the War of 1812, a Representative to the Maine Legislature in 1842 and 1843, and died in 1858. He built another tavern at the corner of Main Street and Elm. During the tavern's final years in business, it wore a sign "Temperance House, J. Mitchell."

Jenks' Tavern was probably the best-known tavern in North Yarmouth. It was constructed by Nathaniel Jenks in the early 1800s and run by him and his son, Captain Seth C. Jenks, for many years. Corliss states that the stage horses were changed there, and travelers could get something to eat and drink. The tavern was a large three story building, similar to the Lane

7 Collection of Yarmouth (Maine) Historical Society.

8 Ibid.

Lane's Tavern—North Yarmouth[9]

Tavern. The Jenks' Tavern burned down on a hot July day in 1856.[10]

The Lane Tavern was built about 1810, by Captain Seth Mitchell and still stands today on Portland Street. Deacon John Webster ran it around 1820, and later Captain Eben Lane, after whom it was named. Judging from its size, entertainment houses were in big demand.[11]

Captain Samuel Larrabee was another tavern keeper in North Yarmouth. "Larrabee served in the Revolutionary Army and upon return married Mary Brown of North Yarmouth in 1778 and settled in his native town.

He built a large tavern on the Topsham Road and became Inn Keeper for several years. Here the old-time stagecoach stopped and changed horses every day—150 years ago. Captain Larrabee was a popular landlord and lived a long life, dying in 1844 in North Yarmouth."[12] Associated with his tavern were barns just across the road. The first elephant which was brought to town was housed in one of the barns in 1816. It brought much excitement to the children of the town.

The "Stage Tavern" was constructed in 1835, by the Maine Stage Company which, no doubt, used it as the stage stop between Falmouth and Freeport.[13] It was located near the Falls, and therefore, must have been near the second bridge across the Royal River which was built

9 Collection of Yarmouth (Maine) Historical Society.

10 William Hutchinson Rowe, *Ancient North Yarmouth and Yarmouth, Maine 1636-1936*.

11 Ibid.

12 Roots Web, Valley Forge Muster-Larrabee. Retrieved August 1, 2017, from RootWeb.Ancestry.com website: <http://archiver.rootsweb.ancestry.com/th/read/LARRABEE/2004-07/1089585782>.

13 Portland Daily Advertiser (Portland, Maine) September 22, 1835.

before 1811. Sam Larabee's Tavern was also located at the Falls, just west of the river. As the Maine Stage Company was winding down its operation in 1844, an ad appeared that the "Stage Tavern" was for sale or to be let. The ad noted that "All the stages on this route stop at this house." There were stables, other out buildings, and three to four acres of land involved.[14]

Alexander Gray may have been the last inn keeper at this location under the Maine Stage Company, as he announced in 1846 that he had moved from his former stand at North Yarmouth Falls to a new place one half mile to the East on the main road to Freeport.[15]

Freeport

The best-known Tavern in Freeport belonged to Samuel Jameson, and still exists today just north of the L.L. Bean store. The tavern was built as a residence in 1779 by Dr. John A. Hyde. He sold it to Captain Samuel Jameson in 1801, who, along with his wife, ran it as a tavern until widow Jamison sold it in 1828. Freeport was a strategic stagecoach town located between Portland and the towns and cities to the north and down east. Such luminaries as the poets Henry Wadsworth Longfellow and John Greenleaf Whittier, as well as the 14th president of the United States, Franklin Pierce, were known to have stopped here on their way to Bowdoin College in Brunswick. The inn was a renowned meeting place for all of the prominent businessmen in the area, and the tap room was well known for its fine spirits and jovial atmosphere. In 1828, Richard Codman purchased the inn and it became known as Codman's Tavern. Codman continued to run it as an inn until 1856.[16]

Jameson Tavern in Freeport[17]

[14] Portland Advertiser (Portland, Maine) June 18, 1844.

[15] Ibid April 14, 1846.

[16] Jameson Tavern Website-History. Retrieved August 1, 2017, from JamesonTavern.com website: <http://www.jamesontavern.com/about-us/history>.

[17] 1915 original postcard published by F. E. Merrill, Freeport, Maine.

Another inn of importance was that of Major Thomas Means, who began his establishment in 1805. Thomas was born in Freeport in December of 1756, shortly after his father and brother, Robert, were killed in an Indian attack at the Means home in Freeport. Thomas Jr. achieved the rank of Major in the Continental Army.[18]

The inn carries an interesting story from those days. "This tale has to do with an Indian who came late one night to the hostelry in search of lodging. Already drunk when he arrived, under the influence of still more liquor, he began to brag of having killed a man. Unfortunately for him, he went on to say that as being one of the raiding party in 1756, he had shot Thomas Means, the Major's father. Apparently, nothing was said or done to indicate the horror his listeners must have felt at this announcement and he was conducted later to his room. There was a small monitor room above the present roof and from this room, he was never seen to depart. In fact, he was never seen again, for the monitor had been the scene of a secret murder. The Indian's ghost is said to have haunted the place until the removal of the monitor room, after which it appeared no more. Before the Major's death some years later, in 1828, he is credited with having told his son that he had made away with the Indian, and in that way, avenged his father's tragic end." Major Means was succeeded by Nathaniel Josselyn, and he by Samuel Bliss. Later, Samuel Thing bought the tavern and continued it for more than 40 years.[19]

[18] *At the Edge of Conflict*, 250th Anniversary of Thomas Means Massacre, Freeport Historical. Society, Exhibit June – November 2006.

[19] Florence G. Thurston and Harmon S. Cross, *Three Centuries of Freeport, Maine.*

Brunswick

Brunswick's first Public house was established in 1740, on the Brunswick Plain, and on the road to North Yarmouth. Robert Spear was the proprietor from 1744 to 1760, and it was not just a tavern, but also served as a garrison house, protected by a timber fortification. Soldiers were kept there to protect travelers, as well as local citizens. Town meetings were held there in cold weather and parishioners went there after church services for lunch and grog. This was another inn which served travelers by horse or foot, as regular stages had not appeared yet.

The next inn, in order of appearance, was kept about 1750 by James Thompson, on the farm at New Meadows. Later, it was kept by his son Brigadier Thompson, who then moved to Topsham in 1784. Other inns were located in Brunswick before 1800, run by Samuel Coombs, Mr. Ross, Benjamin Stone, and Mr. Curtis.

In the first decade of the nineteenth century, John Dunning kept tavern at a place called Washington Hall. David Owen bought it and put on a large addition. He kept tavern there until about 1812, when Isaac Robertson took over.

The Portland *Gazette* carried an ad for Washington Hall in 1811:

> **A Bargain.**
>
> For sale that excellent stand in Brunswick Maine lately occupied by David Owen. One of the best situations for a tavern in this part of the country; about 100 rods from Bowdoin College. The house contains six large and commodious rooms on the lower floor, six convenient chambers, and a

spacious hall; a large stable, sheds and other buildings, which are in excellent order; the garden contains more than an acre of ground. By advancing one third of the purchase money, any person may obtain the stage tavern for much less, than the real value; and the terms of pavement for the remainder will be made easy.[20]

Robertson only stayed three years, selling out to Russell Stoddard, a proprietor in the Maine Stage Company. Stoddard stayed but two years then moved on to the place where the Tontine Hotel would be constructed. Washington Hall then became Ebenezer Nichols' place, until his death in 1824. His widow continued in his place until 1827, calling the tavern "Mrs. Nichol's Inn." Washington Hall burned in 1856.

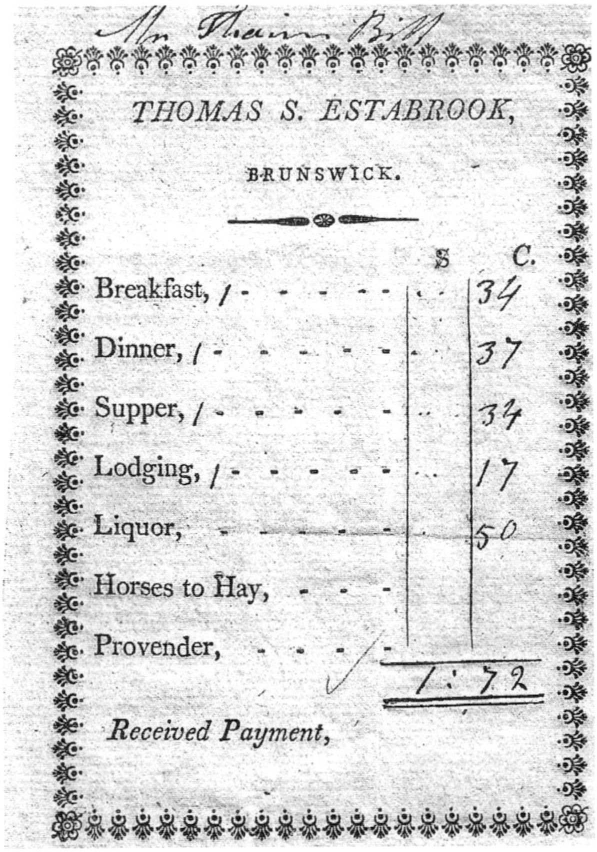

THOMAS S. ESTABROOK,

BRUNSWICK.

	$	C.
Breakfast, /		34
Dinner, /		37
Supper, /		34
Lodging, /		17
Liquor,		50
Horses to Hay,		
Provender,		
		1:72

Received Payment,

A 1790 Bill from Estabrook's Tavern[22]

As one would suspect, a tavern would be located close to or within the campus at Bowdoin College:

> The old tavern which stood in the northwest corner of the college yard, best known as Moorhead's Tavern, was built in 1802 by Ebenezer Nichols. Nichols was landlord of it until 1809. During this time it was the principle public and stage house in the town. Mr. Nichols was succeeded, in 1810, by Colonel Thomas S Estabrook who continued it for some years.[21]
>
> Shortly after taking over the tavern, Estabrook made an arrangement to keep 'Commons' for the tutors and students: The plan was undertaken from motives of economy and convenience. The table was presided over by the Senior Tutor, who said grace. The students were seated alphabetically in order of their classes, and are reported to have been orderly and respectful. The fare, plain, substantial, and inexpensive, was a subject of complaint from a few in each class who had themselves excused from the obligation to board there on the certificate of a Physician that their health demanded an appetite tempted rather than satisfied. After a trial of five years the experiment was discontinued, to be renewed, however, under more favorable circumstances in 1828.[23]

20 Gazette (Portland, Maine May 27, 1811 p4.

21 George Augustus Wheeler, *History of Brunswick, Topsham and Harpswell, Maine.*

22 Courtesy of the Pejepscot Historical Society, Brunswick.

23 George Thomas Little, *General Catalog of Bowdoin College and Medical School 1794.*

After Estabrook left it, a man by the name of Coffin took it and kept it until 1818, when he sold out to Isaac Dow. The latter kept it until 1820, when he committed suicide. In September of this year, it was offered for sale. It was occupied at that time by William Hodgkins. The advertisement stated that it had 'six rooms on the lower floor, some large and well furnished; a spacious and convenient hall, a good cellar and never failing well. The appurtenances are a large, well finish stable, with other outbuildings; and a good garden of more than an acre.'

In 1825, Alexander Moorhead was the proprietor. About the year 1831, Moorhead retired from business, removed from town and engaged in farming. He was succeeded by John L. Seavey, who kept the house for two or three years. The latter was succeeded by James Eliot, who was proprietor until 1839, at which time Mr. Moorhead return to Brunswick and again assumed the management of this public house. Morehead continued to keep it until 1842, when the building was purchased by the trustees of Bowdoin College.

But a short distance west of Cook's Corner, there stood for many years what was known as the Pumpkin Tavern, so-called on account of its sign which was a large ball about the size of a pumpkin, which it greatly resembled. It was a two-storied house built by Wadsworth previous to the war of 1812. Here he kept a public house, about 1836 or 1837. It was seldom patronized by travelers and was in fact more a groggery than a tavern. In the early 1800s, there was a Pumpkin Tavern in nearly every town.

Around 1817, Russel Stoddard opened a public house called Stoddard's, in a large building which stood on the site of the Tontine Hotel. It was built in 1803 for a store, and was occupied as such until Mr. Stoddard took it for a tavern. It was occupied by him until 1825, when it passed into the hands of John Barker and Charles M. Rogers. Barker soon sold out his interests, and Rogers assumed the sole management. The building was destroyed by fire in 1827. During its existence, this house was the stage office and principle public house in Brunswick. It had a hall in which entertainments of various kinds were occasionally held. Stoddard had been in the Washington Hall building for a year or two previous to opening this house.

In 1824, on the corner of Main and Pleasant Streets, William Hodgkins, kept a house known as Hodgkins Inn. A hall, which was situated in the L of the building, was known as Hodgkins' Hall. In 1827, Hodgkins sold out to Charles M. Rogers and moved into the Washington Hall building, which had just been vacated by Mrs. Nichols. Mr. Rogers had been burned out, on the opposite side of the street, where he had been proprietor of the Stoddard House. He remained in his new quarters for about a year when he moved to Topsham, and opened the Lincoln House in November of 1828. He was succeeded in the charge of Hodgkins' Inn by John L. Seavey, who

Pejepscot House and Stables[24]

The Tontine Hotel[25]

[24] Courtesy Pejepscot Historical Society.

[25] George Augustus Wheeler, *History of Brunswick, Topsham, and Harpswell, Maine*.

occupied it as the Maine Hotel until 1830. At this time, it was repaired, refitted, and Major John Stinchfield took it and kept it until 1837, at which time he bought the house of Capitan Schofield and fitted it up for a public house. Mr. James Mustard then became proprietor of the Maine Hotel. In 1840, Mr. Mustard died. The house was then kept for two years by a Erastus Niles, and then for a single year by Joseph W. Sargent, when Mr. Niles again assumed the management and kept it until 1845.

James Mustard, son of the former landlord, took it in 1845 and continued until 1849, when Erin Adams became the proprietor. While under the management of the latter, the house was known as the Pejepscot House. In 1853, a Mr. Marston became proprietor, but he died within a year, and the house passed into the hands and Mr. J. R. W. Hoitt, and was known as the American House.

The Tontine Hotel was built in the year 1828. It opened to the public on New Year's Day, 1829. In a newspaper description, we find: 'It contains thirty apartments besides closets, storerooms, etc., and a hall, which for elegance and spaciousness, is not surpassed in the state.' Colonel Pike was the landlord of the hotel until 1836, when Erastus Richards leased it for two or three years. Jacob M. Berry assumed sole management of the hotel, and continued as its landlord until 1850, when he went to California, leaving the management of the hotel in the hands of Mr. Leonard Townsend. Mr. Berry had been in California only a few months when he died.[27]

It is apparent that basically, the same set of Brunswick inn keepers simply rotated from inn to inn during the first half of the 19th century. This sounds like the coaches with today's professional sports teams!

ACCOMMODATING STAGE

FROM PORTLAND TO BATH.

THE subscribers have established a new line of Accommodating Stages, to run from Portland to Bath, as follows:

Leave Portland, *Mondays, Wednesdays* and *Fridays*, at 10 o'clock A. M. and arrive at Bath the same evening.

RETURNING—Leave Bath *Tuesdays, Thursdays* and *Saturdays*, at 7 o'clock in the morning, and arrive at Portland at one o'clock P. M.

RATES OF FARE.

From Portland to North Yarmouth,	$0 50
" " to Freeport,	75
" " to Brunswick,	1 00
" " to Bath,	1 50

A person has been engaged to drive this Stage who has drove on the above route five or six years, and is a very careful and trustworthy man.

SAMUEL HALE,
I. WATERHOUSE, } *Directors.*

Portland, Aug. 3, 1824.

Bath[26]

Joseph Stockbridge, 1760-1835, was an important inn keeper in Bath. He was born in Abington, Plymouth, Massachusetts, and was a soldier of the revolution, who fought at the siege of Yorktown and served as a sergeant in the corps of light infantry under General Lafayette.[28] In 1785, he

26 Portland Advertiser (Portland, Maine) August 21, 1824 p4.

27 George Augustus Wheeler, *History of Brunswick, Topsham, and Harpswell, Maine.*

28 Parker McCobb Reed, *History of Bath and Environs.*

Bath Hotel -1876[29]

ventured to North Yarmouth and by 1791, had been a mail carrier between Bath and Portland. He married Sarah Mitchell then took his bride to Bath, where they built what was later called the Mansion House.

This tavern must have been in place before 1810, as Governor Gore of Massachusetts had stayed there on a visit to Bath in that year.[30]

The Bath Hotel was built in 1806, by Capitan John Ring. When first erected, its three stories of brick presented an imposing appearance for those days, and made it one of the finest hotels not only of the city, but of the state. It had no piazza; that and the large extension in the rear and its handsome mansard roof being modern additions and improvements. Captain Ring continued as landlord from 1806 through the war of 1812.

When the Bath Hotel was begun in 1806 citizens criticized, as citizens will, the location of the tavern, claiming that it was too far north of the central portion of the city. South Street was then the business hub of our town. Near South Street stood a tavern, the old Shepherd House. Captain Ring, shrewdly foresaw that the extension of the town would be northward, as results proved. The old tavern, which the croakers claimed was too far north, is now south of the center of the city. This incident would indicate the

[29] Courtesy of the Maine Historic Preservation Commission.

[30] Parker McCobb Reed, *History of Bath and Environs.*

> Bath has grown tremendously over three quarters of a century. Captain Ring kept a good old inn at which visitors put up as they journeyed in and out of Bath by the old stage line.
>
> The Bath Hotel was a pleasant, home like hotel, pleasantly situated on Washington street. It was surrounded by the residences of well to do citizens, and it commanded a fine view of the Kennebec River.[31]

This building was apparently the primary stage house in that city, as it was noted that "the Eastern and Western stages will leave the hotel every day in the week except Saturday." Luke Lambard announced in May of 1812, that he had taken over management of the hotel.[32] 1817 found the hotel up for sale "Cheap."[34] Sometimes, with such establishments, a sale occurred, but more often the premises would be leased to a new tenant. In this case, the building's purpose changed dramatically. Mrs. Moody and Miss Low opened a boarding school there, for the education of young ladies. The tuition and board per quarter was only $30![35] Eight years later, Mrs. Adams took it over from Captain Farley to accommodate travelers, transients, and steady borders.[36] In 1827 it was advertised as the Bath House and Stage Office, run by Joseph Blish with the added comment "His table and bar he hopes to furnish so as to meet the approbation of his guests."[37] Within a month, a new steam boat by the name of Experiment came on the scene, plying the waters of the Kennebec River between Bath and Augusta. This gave travelers a smoother way to get to the new Capital, Augusta, and made Bath the transportation hub for points east in Maine. One could travel all the way from Boston to Augusta, or by stage and steam ship from Portland to Augusta. Recognizing a golden opportunity, the passage books for the Experiment were kept at the Bath Hotel.[38] Of course, the Bath Hotel was still the Stage Office, and would be for years to come.[39]

Bath Hotel Trade Card[33]

Less than a year later, with the value greatly increased, the owners tried once again to sell the hotel. If that did not work out, they agreed to lease it out, no doubt at a higher rent.[40]

[31] Bath Independent (Bath, Maine) June 11, 1887.

[32] Eastern Argus (Portland, Maine) May 7, 1812 p4.

[33] Courtesy of a private source.

[34] Gazette (Portland, Maine) June 10, 1817 p3.

[35] Ibid September 2, 1817 p4.

[36] Maine Gazette (Bath, Maine) December 16, 1825 p2.

[37] Ibid May 18, 1827 p3.

[38] Hallowell Gazette (Hallowell, Maine) June 6, 1827. p4.

[39] Maine Gazette (Bath, Maine) September 21, 1827 p4.

[40] Eastern Argus (Portland, Maine) February 19, 1828 p3.

Shepard's Inn-Bath[41]

Thirty-five-year-old Dexter Mellen took over as inn keeper by 1829, but suddenly died on the job in July of that year. A month later, the hotel was being offered for lease once again and then in 1841, the last notice under the name Bath Hotel offers a "recently fitted up" location for lease once more.[42] Following the name changes from Bath, to Exchange, to Phoenix, and finally, Stagecoach Hotel, the building was partially burned on New Year's Day 1964. Two months later it was torn down to make room for a parking lot![43]

> In Bath, as was the case in most cities of any size, taverns were plentiful. One of particular note was that belonging to George Shepard built in 1803. 'Hanging from the corner of the house was a large wooden sign upon which was painted the picture of a traveler dismounting from his horse and over it the words 'Shepard's Inn.'
>
> Shepard's Inn was a favorite place of resort in its day, and the old yellow coaches used to draw up there and let the passengers out for the night, while the horses were taken to the big stables in the rear for rest and feed. The first sight which met the gaze of the traveler as he entered the front hall, was the little window which is still there, and which answered for the bar over which the hot toddies and gin fizzes were served to the guests in exchange for the coin.

41 Courtesy of, Sagadahoc History & Genealogy Room, Patten Free Library, Bath, Maine.

42 Portland Advertiser (Portland, Maine) March 9, 1841 p3.

43 Portland Press Herald (Portland, Maine) January 9, 1964.

> The rooms were all constructed with deep window seats and heavy wooden shutters. A big fireplace was in every room, and a half dozen pair of brass andirons, which have stood as silent sentries for over a century, are still in service looking as bright, attractive, and noncommittal as ever.
>
> The house still holds much of the original furniture, including a mahogany dining table, claw feet and all; several Chippendale bureaus; and other pieces.[44]

It stayed in the Shepard family for many years, and when sold by the last member to reside there, its features were recorded. "The house is colonial in style; woodwork, mantels, even the nails are handmade. The irons from which the sign once hung still project from the outer wall, and in one room is the original wallpaper, brought from Italy by a relative of the builder. The paper is well preserved and shows Italian scenery, with views of the Vatican. This was brought from Italy by Captain John Hall, a relative of the original proprietor of Shepherd's Inn. It is in sheets about a yard square and done in colors. As one enters the room it seems more like the scenery of the theater than wallpaper, and is beautiful and novel beyond description. When one realizes that this paper has been on those same walls for over a century with hardly a scar or blemish, it seems almost incredible.

On the second floor, rooms could be thrown together by raising partitions and hooking them to the ceiling, and in the large apartment thus formed, the Bath Masons held a banquet at the time the Bath Lodge was instituted in 1804. The tavern was located at the northwest corner of Washington and Shepard Streets. The landscape wallpaper now ornaments the Shepard Room in the Boston Museum of Fine Arts."[45]

One more tavern deserves mentioning. James Springer kept the first tavern within the present limits of Bath. James, and his son John, hosted the meetings for the parish church from 1758 until the meeting house was constructed. Both were active in the affairs of the parish. Again, we find a close connection between church and tavern![46]

Woolwich

Although probably not as important as a stage stop compared to Bath or Wiscasset, Woolwich had Day's Ferry across the Kennebec, and a well-known tavern that was conveniently located where the ferry landed on the Woolwich side of the river. The Gould Hathorne Tavern was constructed in 1809 by Robert White, and did provide refreshment to those crossing the river while the stage was preparing for its journey on to Wiscasset. As was found in many taverns of its day, the front hallway had decorative stenciling signed by Moses Eaton, Jr. The original edifice was built by Isaac Smith around 1760, as a store and tavern.

Today the majority of homes, as well as the village of Day's Ferry itself, are on the National Register of Historic Places. The village is the

[44] Bath Independent and Enterprise (Bath, Maine) January 21, 1906.

[45] Henry Wilson Owen, *The Edward Clarence Plummer History of Bath Maine.*

[46] Bath Independent (Bath, Maine) November 25, 1948 pp1&4.

Hathorne Tavern at Day's Ferry Woolwich[47]

starting point for the "Old Stage Road" which follows the course of the Wiscasset-Woolwich Turnpike.[48]

The *History of Woolwich* tells us that:

> Several other men held licenses to operate the ferry until 1830. The Maine legislature in 1830, by an act of incorporation, granted Thomas Stetson of Bath a sole right for two years to operate a horse ferry and other boats at Day's ferry. Stetson's monopoly of the ferry was opposed by four old Woolwhich ferrymen. They were Joseph Day, Henry White, Gould Hathorn, and Peleg Delano. They claimed it was 'an incalculable injury to them.'

Stetson at first had a horse propelled ferry, then later placed a cable across the river for operating the craft. Stetson suggested that the ferrymen on the other side of the river join him in carrying passengers, carriages, cattle, etc. from their side of the river by horse boat. They declined either from lack of finances or because they preferred using old styled gondolas or scows. Eventually his ferry service failed and he sold all his property, land, buildings, horses, wharves, and boats. That ended the Day's Ferry service.

About this time a group of prominent Sagadahoc men incorporated the Sagadahoc Company, which promised more modern means of transportation. They introduced the Sagadahoc, the first steam propelled ferry.

[47] Courtesy Woolwich Historical Society.

[48] Day's Ferry Historic District November 29, 1974.

During the cold winter weather, the stagecoach and the travelers crossed the Kennebec on an ice bridge that froze up river from the usual crossing at Day's Ferry. A story was told of one such trip: 'The driver was concerned about the strength of the ice bridge that day. To lighten the load, he unhitched the horses and sent them across ahead. He had no trouble finding a crew of Days Ferry boys to push the coach and passengers across the ice to the Bath shore. The passengers would have been more anxious had they known that by the next morning the ice bridge had disappeared and the river was again open to the tide.

A proof of the superiority of the steam ferry over the boats at Day's ferry was that it ran in spite of ice. Then too, passengers could stay in a warm cabin while crossing the river. For years there was no fixed schedule, crossing when full but arrangements must have been made to accommodate four-horse mail coaches on the route from Portland to a Rockland prior to the railroad in 1871.[49]

A second tavern was located at Grover's Corner on the Wiscasset side of Woolwich, founded by Benjamin Grover between 1768 and 1789.

Benjamin Grover was a private in the revolutionary war, serving in Colonel Samuel McCobb's regiment during the famed Penobscot Expedition of 1799. This consisted of several dozen vessels and 1000 soldiers and sailors, who attempted to capture the British stronghold Castine. They were defeated by the British fleet and returned home after two months.

Grover's Tavern-Woolwich[50]

Benjamin lived in, owned and operated the public Grover's Tavern, which was on the old post road crossing Woolwich from Day's Ferry to Wiscasset, and on to Rockland. This was the first stop east of the Kennebec for change of horses and hot food and lodging for the weary passengers.

On July 31, 1789 Benjamin's first wife, Lydia, and child and his sister were all drowned in Nequasset Pond as they attempted to cross on a float. Benjamin and his second wife, Joanna, had sixteen children including three sets of twins. The old tavern remained in use until the railroad replaced the stage line in 1871. After that it was the home of the Grover and Bailey families until it was sold to Dr. George Fuller. Soon after it was destroyed by fire in 1939.[51]

49 Woolwich Historical Society, *History of Woolwich, Maine, A Town Remembered.*

50 Fannie S. Chase, *Wiscasset in Pownalborough.* Courtesy of the Wiscasset Public Library.

51 Ibid.

Some notoriety came to the tavern in 1860, as it became the temporary home for nine Eskimos who had been rescued from an ice flow near Labrador. Originally, they were taken to Washington, DC, but the summer heat there made them sick so they moved into the basement at Grover's Tavern, where the coolness made them feel more comfortable.[52]

Wiscasset (Pownalborough)

Taken from the book *Wiscasset in Pownalborough* by Fannie S. Chase.

Bradbury Inn

The house of Josiah Bradbury, inn holder, was designated in the town warrant of 1761, as the place of the first annual town meeting. Bradbury's inn stood on Middle Street, upon land bounded on three of its sides by Fort Hill, Middle, and Front Streets, and comprised lots number 19 and 28. Bradbury was not among the settlors who came here upon the resettlement of this land. Bradbury is described as being a resident of Gloucester, in the county of Essex, and by occupation, a shipwright.

His house was described as '36 feet long and 30 wide, and one story and one-half high, with a brick chimney of five smokes, three below and two above.' 'Little trace of him appears prior to the first town meeting, at which time he was chosen Town Treasurer, to which office he was reelected in 1761, and again elected thereto in 1763, 1764 and 1765.' Of the inn, itself, we have a further description: 'a log house a story and one-half high with the central chimney. The lower story consisted of three rooms, one long kitchen or living room, back of the chimney, and two smaller front rooms. Standing as it did near the shore it was easy of access from the harbor and river, then a great thoroughfare for travelers. To it came the voters of the town to assemble for the transaction of town affairs in town meetings.'

'Although the Bradbury inn was a log house, it bore the distinction of having a larger number of squares of sash glass than any other dwelling at Wiscasset Point, with the exception of Madam Kingsbury's house. Judge Bourne, an accurate writer upon historical subjects, stated the houses were valued then by the squares of glass, or the degree of light which they enjoyed, the number and size of the windows being an important item in fixing their value. From the return of 1766 it would appear that a considerable number of houses here had then, no sash glass.'

In 1811, the town provided for the laying out of a street beginning at High Street and running south easterly to Middle Street. It was given the name of the tavern keeper of long ago, and Bradbury Street is all that is now left to perpetuate his name.

Whittier Tavern and Hilton House

The Whittier Tavern, which became the Hilton House, was one of the first to be designated as a stage house. Actually, they were not the same

[52] 4Grovers.com, for the Grover Family. Retrieved August 1, 2017, from 4Grovers.com website: <http://www.4grovers.com/>.

building, but were sequentially on the same piece of land. In fact, the horses and coaches were housed on the east side of Middle Street, just north of Main Street.

Ebenezer Whittier was born in 1733, in Haverhill, Massachusetts. In 1765, he purchased from Samuel Williamson of Pownalborough for the consideration of four pounds, twelve shillings and eight pence, a certain parcel of land situated in this town, which was described as containing one quarter acre.

Mid-summer of 1766, found Whittier with his family at Wiscasset Point, where two months later, he had raised, upon lot forty-one, the four two story dwelling houses, then standing in the neighborhood. Ebenezer Whittier was admitted an inn holder in 1768, but the Whittier tavern had entertained travelers prior to 1767. Whittier's occupation of joiner enabled him to have built this house with his own hands. Houses, in those days, were built slowly and scrupulously, the wood been given time to season.

The Whittier Tavern stood upon the lot of land where the Hilton House was later built, so the history of both ordinaries can be given together. The Hilton House covered a portion of the exact spot on which the Whittier Tavern stood. The old tavern is said to have been twenty feet longer than the Hilton House and extended more to the eastward. The Whittier Tavern, like the Bradbury Inn, must have had a huge central chimney, for a letter exists from a great-granddaughter of Ebenezer Whittier in which she writes: 'I have often heard my father tell of sitting in the old tavern chimney and looking up at the stars.'

Ebenezer Whittier first appears on the town records under the date of August of 1766, and for the next 44 years, he continued to be a resident of Wiscassett, during which time he held important offices in the government of town and parish. He was a public-spirited citizen and a man of affairs. Ebenezer served the public long and faithfully, both in town and tavern. He was chosen to represent the town in the General Court of Massachusetts, in the year 1787. He was the first postmaster of Wiscassett, in what was said to have been the second post office duly authorized by the Federal government within the limits of the territory of Maine. It was established on February 16, 1790. Among his holdings was part of the old sawmill, located on Mill Creek in the north part of the town, at the site of the former Kincaid Mill. To the Whittier tavern came the post riders John Sutton Foye and Samuel Sevey, both Wiscassett men, bearing the mail on their weekly trips from Portland to Warren, and points further east.

With all of his ramifying interests, it is small wonder that his tavern, conveniently located in the center of town, was the lodestar which drew the village folk within its orbit, and that from it radiated those diverse activities which became an integral part of the town and county. There at the tavern, entertainment was frequently furnished for the members

Hilton House-Wiscasset[53]

> of those two ancient institutions, Lincoln Lodge of Free Masons and Wiscassett Fire Society. The tavern also housed the deliberation of the ambitious wharf and warehouse projects of the proprietors of Wiscassett Union Wharf.

Ms. Chase describes the route into town for the stagecoach: "In later years the stagecoach, after coming through Parker Street, swung around Flagg's Corner and entered the town via sleepy Hollow and High Street. As it approached Bank Hill, the scholars in the bank building would rush to the windows and raise the shades in order to see the coach go by. And when it crossed the common to enter the village, Bover Joe Hilton would announce its arrival with the crack of the whip and a blast of his post horn, six feet long. So heralded, the coach would approach the Whittier Tavern which was sometimes called the Stage House, to be welcomed by no less a personage than Colonel Ebenezer Whittier, its host. The arrival of the stagecoach was the event of the day."

> The Whittier tavern was destroyed by fire on Sunday morning, July 16, 1843. The Wiscassett newspaper printed the following account: 'On Sabbath morning last, about 2:30 o'clock our village was alarmed by the cry of fire, which proceeded from the tavern house of Ebenezer Hilton Esq, and

[53] Fannie S. Chase, *Wiscasset in Pownalborough*. Courtesy of the Wiscasset Public Library.

which had made such progress as to defy all efforts to extinguish it. The spirited exertions of our fire department were very judiciously directed to toward the surrounding buildings, which by seasonable precaution and well directed efforts and the calm state of the weather, escaped uninjured. This fire was fortunately discovered in season to remove most of the furniture of the tavern uninjured, and even the window blinds and some of the windows. Indeed, everything was saved that could be, by the usual activity of our citizens...'

Colonel Ebenezer Hilton, who had married Abigail, the widow of Henry Hoskins and daughter of Ebenezer Whittier, caused a new house to be built that was known by his name. The names of the Hiltons, father and son, are identified with the old stage coaching days, being proprietors of the through stage of the early years. Colonel Hilton was, for many years, a well-known citizen of Wiscassett, he having served in various public capacities. Like his father in law, he was the Postmaster for many years. He was also Selectman in 1823, 1833, and again in 1839 and 1842. His death occurred on September 8, 1843, and his widow, a woman of notable capabilities, maintained an interest in the house and resided there until her death at the advanced age of 84 years. Their daughter, Ann Brooks, married Lewis Hubbard. Mrs. Lewis Hubbard and her sisters, Miss Elizabeth and Miss Susan Hilton, survived their mother and ended their days in the Hilton House.

As a tribute to these three old ladies who for many years took an active part in all of the undertakings of the village, the three stately and enduring elms in their garden were called 'the three sisters.'

Edward Hubbard, who died in 1875, was the landlord of the Hilton House for several years. He was succeeded by his younger brother, William Guild Hubbard, who identified with the hostelry until the Hilton House burned to the ground on the afternoon of October 8, 1903.[54]

Ms. Chase mentions several other taverns, including Sevey's, Foye's, Stacy's, the Wiscasset House, the Washington Hotel, the Pitt Tavern, the Tinkham Tavern, the Nickel's House, the Lee House, Bailey's Tavern, the Marston House and Webber Tavern. Seemingly, all were of local interest, but none were stage houses.[55]

[54] Fannie S. Chase, *Wiscasset in Pownalborough*. Courtesy of the Wiscasset Public Library.

[55] Ibid.

Chapter Nine

Taverns from The King's Highway to Augusta

WISCASSET TO AUGUSTA

Dresden

Travelers from Wiscasset to the Augusta-Hallowell area would proceed up the east side of the Kennebec River, stopping first in Dresden. The newly constructed bridge, over the Kennebec at Augusta, permitted travel by this route. The trip, being only twenty-four miles in length, probably needed only one stop between Wiscasset and Augusta Apparently, the stage house of choice belonged to John Chism, and was located at Dresden Mills. Chism's tavern was a stopping place for stages before the days of the railroad. The sign in front was square lettered, "Dresden House." Chism came from Bowdoinham and Alna to Dresden in 1811, and his original trade was that of a carpenter. After 1840, John Hawthorn and son kept the house. By 1880 or so, it became Sergeant's Inn.[1]

Chism Tavern, Late 1800's[1]

[1] Charles Edwin Allen, *History of Dresden, Maine.*

[2] Courtesy Ann and Jeffrey Pierce.

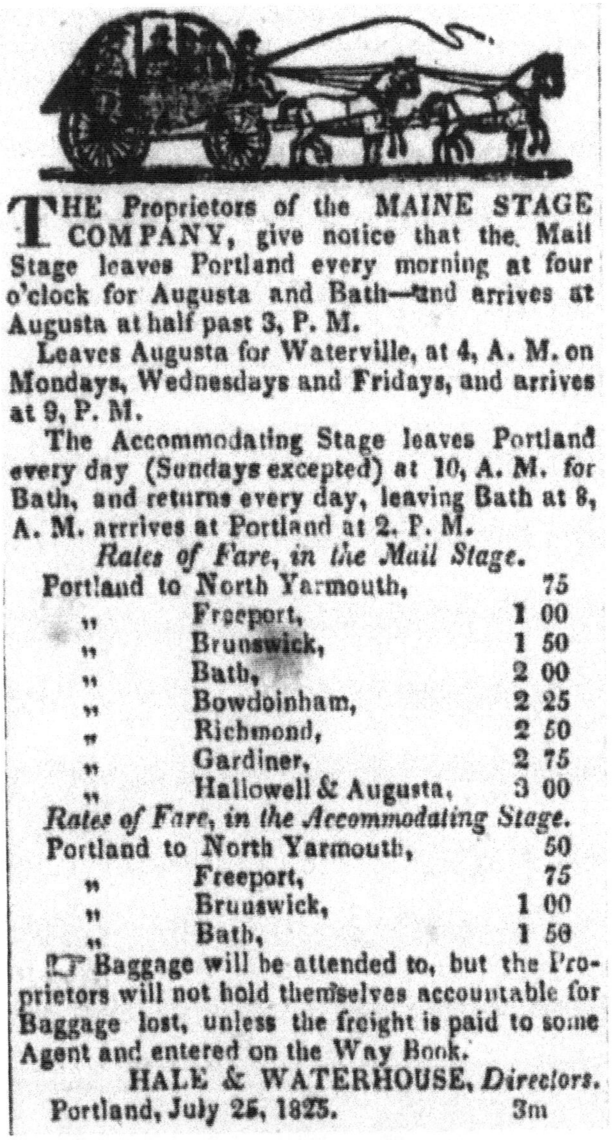

THE Proprietors of the MAINE STAGE COMPANY, give notice that the Mail Stage leaves Portland every morning at four o'clock for Augusta and Bath—and arrives at Augusta at half past 3, P. M.

Leaves Augusta for Waterville, at 4, A. M. on Mondays, Wednesdays and Fridays, and arrives at 9, P. M.

The Accommodating Stage leaves Portland every day (Sundays excepted) at 10, A. M. for Bath, and returns every day, leaving Bath at 8, A. M. arrrives at Portland at 2. P. M.

Rates of Fare, in the Mail Stage.

Portland to	North Yarmouth,	75
"	Freeport,	1 00
"	Brunswick,	1 50
"	Bath,	2 00
"	Bowdoinham,	2 25
"	Richmond,	2 50
"	Gardiner,	2 75
"	Hallowell & Augusta,	3 00

Rates of Fare, in the Accommodating Stage.

Portland to	North Yarmouth,	50
"	Freeport,	75
"	Brunswick,	1 00
"	Bath,	1 50

☞ Baggage will be attended to, but the Proprietors will not hold themselves accountable for Baggage lost, unless the freight is paid to some Agent and entered on the Way Book.

HALE & WATERHOUSE, *Directors.*

Portland, July 25, 1825. 3m

Bath to Augusta[3]

Richmond

Peacock Tavern

A canary yellow stagecoach, one of the first stagecoaches used in Maine, was a common sight at the old Peacock Tavern on the Post Road outside of Gardner. Nine inside and six outside was the usual passenger capacity, although on occasions, it was possible to seat four or five on top. Sometimes two horses, sometimes four, and sometimes three, with one as a leader, drew up in front of the inn with a flourish. Some of the story of the Peacock Tavern has been lost in the past, but if it were pieced together, it would probably read something like this: In 1799 Benjamin Shaw, father of the Benjamin Shaw who was later prominent in Richmond history, bought a little store on the shore of Pleasant Pound. The store had belonged to a man named Perkins, who located there as most of the travel was by water rather than land. With the store was plenty of land and Perkins, sensing a change in travel conditions, started a frame house away from the pond and near what became the main road to the growing communities of Central Maine. It was the second building that Shaw bought and in 1803, he settled there with his family.

The house which, originally consisted of one room, was constantly enlarged in a continual building process that must have gone on for some years. That first room is now on the south side and in the back of the inn. A tremendous fireplace, and doors with a Christian cross in the upper half, show plainly that this is the oldest part of the building. Eventually, the house was finished in a perfect square with a central door with sidelights and a rather ornate design on top. Squares within squares formed the ornamentation of the door, and the same motif is repeated on an upstairs

[3] Eastern Argus (Portland, Maine) August 2, 1825 p4.

mantle piece, except that the squares have been turned so that any angles, come at the top and bottom. Obviously, the front door and those upstairs rooms were constructed about the same period.

Later on, probably from about 1830 to 1840, a third segment was added to the house on the north, almost a third of the original size. The second floor, which is now four rooms, was a large dance hall with a fireplace in the center. An unusual feature of two of the second-floor rooms is a fluted chair rail, which looks as if it were accordion pleated. That same design is repeated in two metal pieces. Apparently, the designer had an eye for fancy details.

It was under the management of the second Benjamin Shaw that Peacock Tavern reached the height of its prosperity. The room which was the bar boasted a large fireplace, which is still there. The dance hall, which was built during his regime, was a popular spot and a bowling alley was an additional attraction. Squire Shaw, as he was locally called, was elected chairman of the first Richmond Board of Selectman and the first town meeting was held in the tavern.

Past the old inn came not only the daily stagecoaches, but also a continuous procession of odd looking Canadian travelers bound for the Harpswell fishing grounds. Strange sleds with runners and fills of one piece of wood were a common sight. The road was a lively highway, and the inn a welter of activity before the advent of the railroad. After that, freight and passenger travel died down, but not before Edward Peacock had given the inn the title which it has kept until this day.[4]

Gardiner

Much like Topsham, Gardiner was located too close to the larger towns of Hallowell and Augusta to merit a stage house. By stopping at the Peacock Tavern in Richmond, the total distance to Augusta was just twelve miles, well within the reach of a single span of horses. As with other towns in Maine, there were a few taverns of note. Frequently taverns were used for public gatherings such as public auctions. One such auction occurred at Benjamin Randall's Tavern in Gardiner in 1810.[5] Another took place in 1825, at McLellan's Tavern in that city.[6] As early as 1803, Mr. Shaw also ran a tavern in Gardiner.[7]

BRUNSWICK TO AUGUSTA

Topsham

Although there were several taverns in Topsham over the period of 1770 to 1845, it is doubtful that a stage house was among them. Topsham's proximity to Brunswick, from which the stage route began, precluded the need to change horses after just two miles' travel. We do know that Nathaniel Green was a licensed inn holder from 1814 to 1829. He also kept a public house from 1831 to 1836, after which he went to Augusta

4 Kennebec Journal Unknown Date.

5 Herald of Liberty (Augusta, Maine) June 12, 1810 p3.

6 American Advocate (Hallowell, Maine) August 13, 1825 p4.

7 Kennebec Gazette (Augusta, Maine) September 15, 1803 p4.

Portland & Augusta Stage Line.

THE subscribers hereby give notice that they have made an addition to their Stage line from Portland to Augusta, which will run as follows: Leave Portland every day at 6 o'clock A. M. run through Falmouth, Cumberland, N. Yarmouth, Freeport, Brunswick, Topsham, Bowdoinham, Gardiner, Hallowell, and arrive at Augusta the same evening. Returning, leave Augusta every day at 4 o'clock A. M and arrive at Portland the same evening. And they pledge themselves to furnish good horses, carriages, and careful drivers, and hope to receive a share of public patronage.

S. HALE,
R. STODDARD,
I. WATERHOUSE.

Portland, Dec. 28, 1822

Brunswick to Augusta[8]

to manage the Palmer house. A public house located on Winter Street was kept by Sullivan Haynes prior to 1826, and by Prince Dinsmore after that date. Later that year, it burned down and an interesting story was recorded:

> **Fire in Topsham Again**
>
> On Saturday evening, last, between the hours of nine and eleven the house owned by Captain Samuel Perkins of Topsham and occupied by P. Dinsmore as a tavern, situated in the center of Topsham village, together with a barn and other out houses, were destroyed by fire. Mr. Dinsmore lost a considerable part of his furniture, etc. The buildings were worth about $800. No part was insured. Much credit is due to those Brunswick people, who assisted in the very dangerous undertaking of laying plank upon the bare stringers of the bridge, (which had closed for repairs with an open gap of about 50 feet) in order to convey the Brunswick Engine across the same. This was the more creditable, as the Topsham Engine became useless through a defect in the hose immediately after they began to play water upon the fire, and much more damage would probably have been done but for the timely assistance of the Brunswick Engine. No cause can be assigned for the origin of this fire.[9]

The house was rebuilt in 1828 by Charles M. Rogers of Brunswick, who advertised it as the Lincoln Hotel, "A new and commodious house." Interestingly, Roger's tavern in Brunswick had burned down the year before. "Mr. Rogers had been burned out in his tavern where he had been proprietor of the Stoddard House. He remained in his new quarters for about a year, when he moved to Topsham and opened the Lincoln House in November, 1828."[10]

> James Mustard ran the Lincoln Hotel from 1830 to 1834, and Jeremiah Clough kept it in 1837. In between, it was kept by Suel and Ablen Baker who renamed it the 'Temperance Hotel.'[11]
>
> About 1770, Mr. John Hunter kept a tavern about two miles from the village, on the road to Bowdoinham. Nothing definite is known as to the length of time Mr. Hunter kept it as an inn, but it was probably for five or six

8 Eastern Argus (Portland, Maine) January 14, 1823 p1.

9 Eastern Argus (Portland, Maine) July 14, 1826 p3.

10 George Augustas Wheeler, *History of Brunswick, Topsham and Harpswell.*

11 Ibid.

John Hunter Tavern (The Old Red House) in Topsham[12]

years. He was Town Clerk from 1773 to 1775. He died at 32 years of age. In 1777, a town meeting was held at 'Widow Hunter's.' From these facts it is probable that Mr. Hunter died in 1775 or 1776, and he had probably kept tavern for some years previous to that time. Mrs. Hunter carried on the business for some years, until she married Mr. Alexander Rogers. It used to be a custom for parties of five or six to ride from the village to this inn, and for the last one who arrived to pay for the 'treat.' Town meetings were occasionally held at this house, when the severity of the cold made the meeting house too uncomfortable.

While Mrs. Hunter carried on this house, an old soldier named Pike (pronounced Peck), returning from the war in tattered clothes and with his faithful musket on his shoulder, begged of her to allow him to remain and work upon her farm. She consented and gave him plenty to eat and a new set of clothes; whereupon, he remarked that he would stay as long as he lived. In after years he remarked in still stronger terms, that he would remain with her 'as long as a single shingle remained on the roof.' The old house still stands in its oaken strength, while Pike, a faithful servant, has long since moldered in the dust. After Widow Hunter's marriage, the house ceased to be a tavern. However,

[12] George Augustas Wheeler, *History of Brunswick, Topsham and Harpswell.*

Mr. Rogers in 1803, and for some years after his marriage to Mrs. Hunter, kept an inn at his own house, and it was the resort for lunch and grog on Sundays for all who attended meeting in the old east meeting- house. Here too, the militia collected on training days, and here the procession was formed when Washington's death was observed, in 1800. The house descended from Mr. Rogers to his son, honorable George Rogers, and from him to the late George A. Rogers, Esq.[13]

Bowdoinham

Mustard Tavern

Here, the first stop was at Bowdoinham and the Mustard Tavern.

> Joseph Mustard came to Bowdoinham in 1819 and came in possession of the inn soon after running it as a public house until 1853 when it ceased business. The road from Zebulon Jack's to Richmond Corner and the Pond was built in 1807 so the hotel was erected 10 years before the road was built.
>
> There might have been quite a lot of travel in that direction, enough to induce the proprietor to build, but it must have been wholly by horseback. The settlement of Litchfield and other sections in the vicinity gave the prospect for an opening which could not have been very flattering.
>
> From information given, Joseph Mustard, Sr., came originally from Woolwich with his son Joseph, Jr., he 'swapped' the place with Isaac Leonard in 1819, and four years thereafter Mustard's Tavern was one of the best known public houses in this part of the country. It is today called by its old name, 'Mustard's Tavern,' and so long as it stands, and it looks good for another century at least, it will still carry the title which it so well earned. Being upon a direct stage line between Augusta and Portland, it could not well help becoming prominent, and here it was that all stages stopped. For a change of horses and to secure refreshments for the travelers. The big yellow coaches dashing along over the roads, drawn by four horses did a large business before and the iron horse came into use, and to supply fresh animals for these coaches there were always seventeen to thirty-five horses stabled in the old barn across the road which is still standing, although it was built over one-hundred years ago.
>
> This ancient hotel, once the scene of so much gaiety and bustle, has been kept in a good state of repair and in the front yard loom those plants so popular in old New England's gardens of a century ago. The interior of the old tavern has undergone no great changes since the days when it was a public house. Connected with a tavern were large fields and gardens wherein were raised the vegetables for the table and the grain and hay for the stock. It is said that Leonard Mustard

[13] George Augustas Wheeler, *History of Brunswick, Topsham and Harpswell.*

used to raise 1000 bushels of potatoes in a season, 200 bushels of shelled corn and other vegetables and grains in like proportions.[14]

FREEPORT TO AUGUSTA

The Middle Road

The exact route taken from Freeport to Hallowell and Augusta on the "Middle Road" is not clear. The ad seems to distinguish between this route and the route through Brunswick. By using the name Middle Road, are we to assume that this route broke off halfway between Portland and Bath to head due north through Durham, Lisbon, and Litchfield as the Center Route would proceed at a later date? If this assumption is correct, then the additional stage houses and taverns in those towns will be covered under Other Routes, Chapter 10. With the distance on state routes 125 and 136 being just nine miles, there would probably be no need to stop between Freeport and Durham.

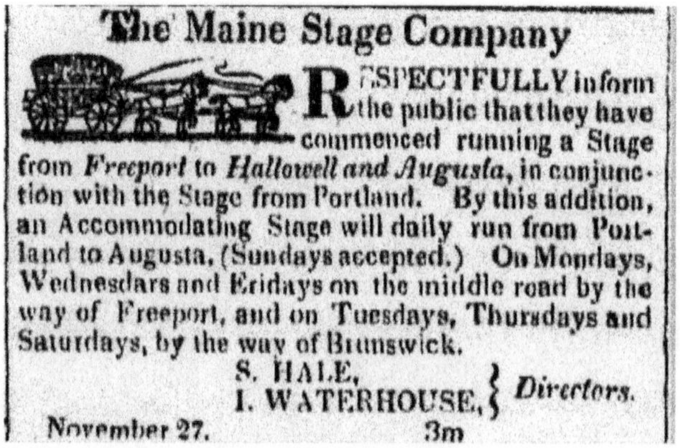

The Maine Stage Company

RESPECTFULLY inform the public that they have commenced running a Stage from *Freeport* to *Hallowell and Augusta*, in conjunction with the Stage from Portland. By this addition, an Accommodating Stage will daily run from Portland to Augusta, (Sundays accepted.) On Mondays, Wednesdays and Fridays on the middle road by the way of Freeport, and on Tuesdays, Thursdays and Saturdays, by the way of Brunswick.

S. HALE,
I. WATERHOUSE, } *Directors.*

November 27. 3m

Freeport to Augusta[15]

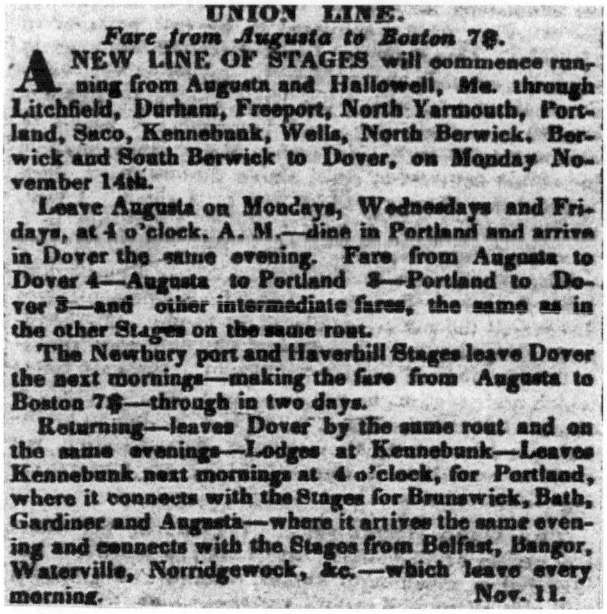

UNION LINE.

Fare from Augusta to Boston 7$.

A NEW LINE OF STAGES will commence running from Augusta and Hallowell, Me. through Litchfield, Durham, Freeport, North Yarmouth, Portland, Saco, Kennebunk, Wells, North Berwick, Berwick and South Berwick to Dover, on Monday November 14th.

Leave Augusta on Mondays, Wednesdays and Fridays, at 4 o'clock, A. M.—dine in Portland and arrive in Dover the same evening. Fare from Augusta to Dover 4—Augusta to Portland 3—Portland to Dover 3—and other intermediate fares, the same as in the other Stages on the same rout.

The Newbury port and Haverbill Stages leave Dover the next mornings—making the fare from Augusta to Boston 7$—through in two days.

Returning—leaves Dover by the same rout and on the same evenings—Lodges at Kennebunk—Leaves Kennebunk next mornings at 4 o'clock, for Portland, where it connects with the Stages for Brunswick, Bath, Gardiner and Augusta—where it arrives the same evening and connects with the Stages from Belfast, Bangor, Waterville, Norridgewock, &c.—which leave every morning. Nov. 11.

Union Line Advertisement[16]

Another ad, by a competing stage line, seems to clear up the question. The Union Line clearly states that the route from Augusta to Portland went through Litchfield, Durham, Freeport, North Yarmouth, Portland, and on to Dover.

Augusta

Certain that Maine had become an independent state in 1820, but unsure of the future location of the Capital, Augusta continued to grow and hope. Travel by stage had started as early as 1807, but it was not until 1826, with the advent of the Maine Stage Company that Augusta's position as the transportation hub of Central Maine was secured. A year later, Augusta became the capital of Maine.

Two of the principal stage houses in Augusta appeared on the same advertising page of the *Subscribers Business Directory* for Maine in 1858. Although this was fifteen years after the railroads had replaced most of the stage traffic,

[14] Silas Adams, *Town of Bowdoinham 1762-1912.*

[15] Eastern Argus (Portland, Maine) February 19, 1828. p4.

[16] Eastern Argus (Portland, Maine) November 22, 1831 p1.

Augusta in 1823[17]

it does give us a good idea of the size and importance of these two establishments.[18]

From a letter written by Harry E. Burbank, in 1894, to a Mr. Stinson who was planning to write an article on the early hotel keepers of Augusta, we learn the following.

Burbank's father had lived in Augusta for three years from 1831 until 1833. There he had worked for Colonel Pettingill who was a large property owner.

At that time, there were seven taverns or public houses in Augusta. Augusta was a great center for the different stages which ran into the different parts of the state, and it required a number of hotels or taverns to keep the many country people who came to Augusta to purchase their goods and supplies.

On State Street, one could find the Augusta House, the Cushnoc House, and the Mansion House. On Water Street were the Kennebec House, the Franklin House, and the Piper Tavern. Roger's Tavern on Cushnoc Heights completed the group. Cushnoc was the original name for Augusta.

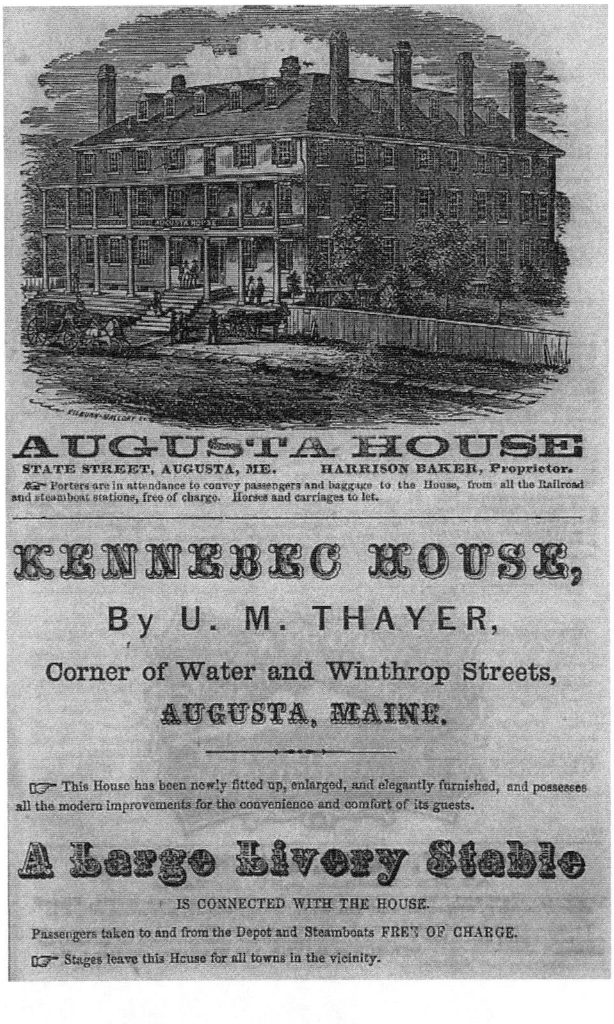

17 Drawing by Searle from James W. North's *The History of Augusta*.

18 *Business Directory of the Subscribers to the New Map of Maine*, 1858.

Kennebec Tavern

The Kennebec Tavern, sometimes known as the Kennebec House, was the first on the scene, beginning in 1784. The first tavern keeper recorded was Amos Pollard, who kept the house until 1797, turning it over to Peter Voss who gave it the name Kennebec House.[19] By 1810, it went by the name of Kennebec Tavern and was kept by T. Hamlen.[20] That establishment is not heard of again until 1821, when T. Hamlen announces that it is once again in his possession and he has plans to "repair and furnish it in the best manner for a Public House of Entertainment."[21]

By 1825, it was called the Kennebec Tavern and Augusta Stage House, Maine. T. Hamlen is still in charge and he notes that: "No less than sixteen stages arrive and the same number depart every week, besides the steamboat line from Bath." To this he adds: "Also a constant supply of horses and carriages on hand for the private conveyance of those who cannot obtain it in the stages and steamboats."[22]

A new stage line was introduced in 1827, running between Belfast and Augusta. The stage books were kept at Hamlen's Kennebec Tavern in Augusta, and at Appleton's Hotel in Belfast.[23] With the advent of the Kennebec Steamboat Line in 1828, sailing from Bath to Augusta via Richmond, Gardiner, and Hallowell, an alternative to stage travel, was available. It still took the same six hours to cover the distance, but the ride had to be much more comfortable.[24] Theophilus Hamlen tried to sell his tavern in 1829. The ad reads:

> **Kennebec Tavern for Sale.**
> "The subscriber offers for sale the above well-known tavern stand and stage house in Augusta; the house where all the principle stages arrive, and where the stage and steam boat books are kept. It is on the lower street, in the center of business, in the immediate vicinity of the Steam Boat Wharf; has been a tavern for forty-five years, and is decidedly the most valuable stand in the interior of Maine. Augusta having been made the capital of the state, this property is daily enhancing in value. An excellent bargain will be offered, a long credit given if requested, and possession given on the first day of October next."[25]

Apparently, G. W. Perkins must have leased the property from Hamlen for a year or so because in 1830, W. F. Brown announced that he had taken over the Kennebec Tavern on the south side of Market Square and Water Street, from Perkins. Brown states that: "Its location renders it very convenient for the gentleman of business, and pleasant for the traveler; and from the improvements which have been added, and the attention which will be paid to his customers, the subscriber feels justified in anticipating a share of the public patronage. Books of all the stages arriving in this town are kept at the bar."[26]

19 James W. North, *The History of Augusta.*

20 Herald of Liberty (Augusta, Maine) July 10, 1810 p3.

21 American Advocate (Hallowell, Maine) December 8, 1821 p4.

22 Boston Traveler (Boston, Massachusetts) August 23, 1825 p3.

23 Eastport Sentinel (Eastport, Maine) February 10, 1827 p3.

24 Eastern Argus (Portland, Maine) May 23, 1828 p3.

25 Bangor Weekly Register (Bangor, Maine) June 9, 1829 p3.

26 Maine Patriot and State Gazette (Augusta, Maine) May 5, 1830 p4.

The year 1832 has Hamlen trying to sell the tavern and associated stables once again.[27] The tavern continued in business for many years thereafter, at least up until 1858, when it appears in the previously noted advertisement.

A side story which well depicts the problem with wooden buildings in the nineteenth century is told in a Portland newspaper in 1841.

Destructive Fire

Last night, between 8:00 and 9:00 a fire broke out in a large wooden building on Water Street, in this town [Augusta,] owned by Colonel William Bailey and occupied by him for storing and pressing hay. There was a large quantity of hay in it, and it consequently made a prodigious flame, which was bursting out in all directions before the alarm was given, and before any engine was put into operation. The fire had communicated to several other buildings on the same side, and two buildings on the opposite side of the street. The large brick store built by Charles Williams, and occupied by Wheeler and Perkins, and others, proved a successful barrier to it to the north (the wind being southerly and moderate,) but the fire spread to the south, consuming the building occupied by Morse and Jayne, Augustus Brick and others, and also a large old dwelling house adjoining, and we believe one or two small ones.

On the opposite side, too, the building known as the Old Heath House Tavern, and occupied by L. M. Judkins, the house of James Black, the meat store of Mr. Page and the blacksmith shop on the corner, were all consumed. The house of Mrs. Prescott and part of the stables of the Heath House were saved. We did not learn last night who of the sufferers were insured, nor the amount of their losses. We regret to add that Mr. Simeon Goodwin was badly hurt by falling from the roof of the Kennebec Tavern, whither he had gone to extinguish the fire on the roof. Perhaps twenty other buildings would have taken fire from falling cinders, had not the roofs been kept wet by water carried up and by wet blankets.[28]

Augusta House

The Augusta House was certainly in place before 1828, as a public auction was held there in that year to sell timber lands.[29] James North, in his book *The History of Augusta*, tells us that it was then owned by Major Harrison Baker who sold it to some businessmen and the City of Augusta to be enlarged and remodeled. The new edifice was built of brick in 1831 at an expense of $26,000. The proprietors incorporated in 1831 and the house was taken over by Nathaniel Green in 1832, who we remember from Topsham.[30]

The hotel was up for lease again in 1837 following the death of Mr. Levi Rogers who had taken over from Mr. Thomas Stevens. The advertisement states that "Connected with the hotel are extensive stables, in which have been accommodated most of the lines of stages running into Augusta."[31] It also notes that it has a commanding view of the Kennebec River, the

27 Age (Augusta, Maine) October 3, 1832 p1.

28 Christian Mirror (Portland, Maine) April 1, 1841 p3.

29 Maine Patriot and State Gazette (Augusta, Maine) September 10, 1828 p4.

30 Portland Advertiser (Portland, Maine) October 29, 1832 p3.

31 Age (Augusta, Maine) August 2, 1837 p1.

United States Arsenal, the Insane Asylum, and the village of Augusta. In the fall of that year, Mr. Hutchins took over and management tried again to address its original purpose by soliciting members and visitors of the Legislature, along with transient visitors during the summer season.[32] Having previously owned it, Major Harrison Baker ran it in 1858, and leased it again in 1865 and in 1870. It contained 120 rooms and graced the skyline of Augusta for many years. A third stage house came on the scene by 1836.

Mansion House

In that year John Barker and Henry Hobbs notified the public that they had taken the spacious establishment in Augusta known as the Mansion House. They included the statement that: "Stage books in all the lines east, west, north and south kept here."[33] Later that same year, an article from the Boston Traveler was copied by the Age Newspaper of Augusta, entitled "Mansion House, Augusta."

> One of the noblest publicans in Maine, is John Barker, who keeps the Mansion House in Augusta, recently conducted by Levi Rogers. His fame had reached our ears, we had heard his merits disclosed in stages, steamboats and hotels, by those who had participated in his good cheer and elegant accommodations; and during a late visit to the Capital of our sister State, we had opportunity to test the truth of the current eulogiums. High as were our expectations, they were fully realized in a brief sojourn at the Mansion. We found 'mine host' all he had been represented, active, liberal, attentive and polite; his table literally loaded with the solids and luxury of the best markets; the attendants numerous and civil; the apartments convenient and handsomely furnished, in fine, the whole in a style of elegance and comfort, that would sustain the reputation of Stetson of the Tremont, and Holland of Nahant. Barker took the Mansion House in January last, and has thus far been patronized to the extent of his hopes. While some of the Eastern Hotels have languished for business this summer, his numerous apartments have been filled, and at times applicants were reluctantly compelled to seek lodgings elsewhere. As an evidence of this enterprise, and his determination to excel in the 'means and appliances ' for a first-rate table, it may be mentioned that he last winter kept one or two men constantly employed in scouring the country for the choicest steak, poultry and wild game.
>
> To accommodate all his friends, Mr. Barker is now making a large addition to his establishment which, when completed, as it will be in a few days, will give 100 feet front, with a piazza extending the whole length. It will then be one of the most elegant and commodious public houses in the country. Connected with the hotel is a large stable, where visitors may procure the finest horses for a drive that can be obtained anywhere; and one, at least, will trot to Hallowell and back, four miles over an uneven road

[32] Ibid December 6, 1837.

[33] Ibid February 10, 1836 p1.

> in sixteen minutes. Everything, indeed, connected with the Mansion House, goes ahead in like manner.[34]

The next year, a stage stable one-hundred-forty feet in length, was added at an estimated cost of three to four thousand dollars.[35]

More kudos for Mr. Barker appeared later that summer.

> The beautiful village of Augusta, Maine, possesses more interesting objects, whether as regards natural scenery, or elegant public edifices, then any town of its size in New England. These alone would be sufficient to make it pleasantly remembered; but we suspect that the village is still more indebted for the enviable place which it holds among the reminiscences of every visitant, to the unrivaled professional merits of Mr. Barker, keeper of the Mansion House. Uniting the courtesy of a gentleman to the frank hospitality of a country landlord, he produces an impression which is equally delighted and unique. His guests are safe from every disagreeable emotion, except such as are excited at the moment of their departure; for, go where they may, they can never hope to find softer pillows, or a board more richly spread, or champagne with a brighter sparkle, or a more generous welcome to the whole, than beneath the roof of Mr. Barker.[36]

[34] Age (Augusta, Maine) September 21, 1836 p3.

[35] Gospel Banner (Augusta, Maine) June 10, 1837 p3.

[36] Gospel Banner (Augusta, Maine) July 22, 1837 p3.

The Maine Stage Company announced in September of 1837 that they were moving the stage books for Augusta to The Mansion House. Although that would be the new stage house for Augusta, they would still agree to pick up or set down passengers at any preferred location.[37] Mr. Barker must have been a very persuasive inn keeper! Less than a year later Barker had switched from the Mansion House in Augusta to the Cumberland House in Portland.[38]

A Mr. Stevens kept the house until 1842 at which time Stephen Scruton was the manager, coming from the New Hampshire Hotel in Dover. Stages arriving from different parts of the country still stopped there and the books for the same stayed there.[39]

Cushnoc House

This was the oldest hotel in the city, built around 1801. It was run by a man named Nickerson early on, and then was taken over by Mr. Dillingham, whose son was a Universalist minister. By 1839, Mr. Hobbs had taken over the management of the house from Johnson and Moore. In 1845, after extensive repair and refitting, A. H. Scruton assumed control. The house was named after the original Indian name for Augusta.

Piper Tavern

Giving up his trade as house wright or carpenter, Benjamin Piper, who came to Maine from Pembroke, New Hampshire in 1801, went

[37] Age (Augusta, Maine) September 6, 1837 p4.

[38] Portland Advertiser (Portland, Maine) May 1, 1838 p4.

[39] Ibid.

into the tavern business. For many years he conducted the brick tavern at the corner of Laurel and Water Streets.[40] It was built using bricks from a local brickyard near present day Monument Park. Years later, it was known as the Farmer's Hotel. During the Civil War, it was focal point for organizing troops such as the 7th Maine Regiment. Later on, it was run by John Fuller, Esq.

Franklin House

A house by this name is advertised in 1842. At that time, J. P. Philbrook invited members of the Whig, Democratic, Federalist, Conservative, and even Abstractionist parties to stay at his inn. He also notes: "To members of the Legislature. The subscriber would respectfully give notice that the Franklin House affords as comfortable winter quarters as any public house in town, and at prices so reasonable that a reasonable man cannot object."[41]

Following Phillbrook, Nathaniel Hutchins, previously of the Augusta House, took over as manager in 1844. He noted that the house was centrally located in the business part of town and equipped with excellent stables and yards.[42]

Roger's Tavern

Mr. Rogers was also in the lumber business, and was known as "Wheelbarrow" as he marked all his logs in the river with a wheelbarrow impression to distinguish his property from that of his fellow lumbermen. This was a great resort for country people who would make this their headquarters while making their purchases in the city. John Fuller also ran this place for a number of years prior to his running the Farmers Hotel. In those days, all the hotels sold liquor openly, it being then largely styled under the heading of "West Indies Goods."[43] This tavern still stands on Northern Street.

Reed's Tavern

Another existing original tavern (now a private residence) is located on Bangor Street. In 1836, when a plan of the building was drawn by B. F. Perham, it was known as Whitney's Tavern. From its location, it was popular with travelers to Bangor and points east.[44]

Fuller Tavern

In the heyday of the ten or more taverns in Augusta, there was none more favored than the Fuller Tavern on Maintop, built and kept by John J. Fuller early in 1800. It was popular with travelers from the country north of Augusta. A later owner was Joel Richardson, who sold it in 1867 for $10,000, a fabulous sum in that day.

In the attic were four chambers for overnight guest, while the front hall was very large. It represented a good example of early Augusta tavern architecture.[45]

40 Kennebec Journal (Augusta, Maine) February 22, 1932.

41 Age (Augusta, Maine) May 6, 1842 p5.

42 Maine Farmer (Augusta, Maine) July 25, 1844 p3.

43 Letter from Harry E. Burbank to Mr. Stinson courtesy of Anthony Douin, Kennebec Historical Society.

44 Kennebec Journal (Augusta, Maine) March 28, 1932.

45 Undated and untitled newspaper article, Kennebec Historical Society.

Chapter Ten

Other Routes to Augusta and Hallowell

New Mail Route,
FROM PORTLAND TO HALLOWELL.

The Maine Stage Company, desirous of affording greater accommodation to the public than they at present possess, have established a line of

STAGES,

Between Portland and Hallowell,

by the way of Falmouth, Pownal, Durham, Lisbon and Litchfield, to run twice a week.

The Stage will leave Portland every Tuesday and Friday, at 7 o'clock, A. M. and arrive at Hallowell the same evening.

And will leave Hallowell every Wednesday and Saturday, at 7 o'clock, A. M. and arrives at Portland the same evening.

The Proprietors have determined to keep upon this route, good Coaches, fleet horses, and careful and trusty drivers, and hope to receive the patronage of the public.

The Stage Books will be kept at Bowles' *New Hotel*, in Portland, and at Dillingham's *Hotel*, in Hallowell.

SAMUEL HALE, ISRAEL WATERHOUSE, } *Directors.*

January 12, 1826. 2

New Route to Hallowell[1]

Although North Yarmouth is not included in the above listed stops for the stage line, it is quite probable that Maine State Route 9 was followed out of Falmouth, through Cumberland and Walnut Hill, in the corner of North Yarmouth. Using that assumption, the next stop after Falmouth would have been at the Buxton Tavern which is right on the Yarmouth-Cumberland line.

For convenience, we shall call this the "Center Route." It was probably the most direct route to the Kennebec Valley.

1781 was the year that Isaiah and Sarah Cushman built the famous "Red House" on a sixty-five-acre lot in Walnut Hill. Jeremiah Buxton bought it in 1795 after the county road from Yarmouth Village to New Gloucester had been established. A few years later, the road to Hallowell would intersect the county road right in front of the Buxton place. The Buxtons became well known from their business skills.

Jeremiah married Jane Drinkwater in 1796, and the entrepreneurial couple went to work. They established a tavern, farmed, raised cattle, and

1 American Advocate (Hallowell, Maine) January 28, 1826.

Buxton Tavern-North Yarmouth[2]

William Buxton[3]

traded. The oldest son, William, pitched in and followed his parents lead, in time owning shares in the Union Wharf in Yarmouth, the stagecoach which ran past the tavern, and later on the railroad.

Of the six surviving children, Jeremiah Jr. ran the Red House Tavern, while Edward ran the tavern stand located directly across the road. In addition, at the intersection, the family ran a store, a blacksmith shop, and cooperage. Stagecoach travelers stayed in the "Red House", while cattle drovers and peddlers stayed at the stage stand across the street.

On Jeremiah's death in 1835, the "Red House" passed on to his son William, and the family eventually sold it in 1881, after 100 years of occupancy. The most intriguing story about this old tavern concerns the hidden pantries accessible from a false back behind a very shallow cabinet. Were these hiding places for

2 Courtesy North Yarmouth Historical Society.
3 W. W. Clayton, *History of Cumberland County, Maine.*

the family in case of an eminent danger, or simply dead space resulting from the fireplace arrangement? Another prevalent version says that they were used to conceal runaway slaves en route to Canada. There was even a suggestion that there might have been a tunnel connecting the tavern and the stage stand across the street. No physical evidence has been found for the latter supposition.[4]

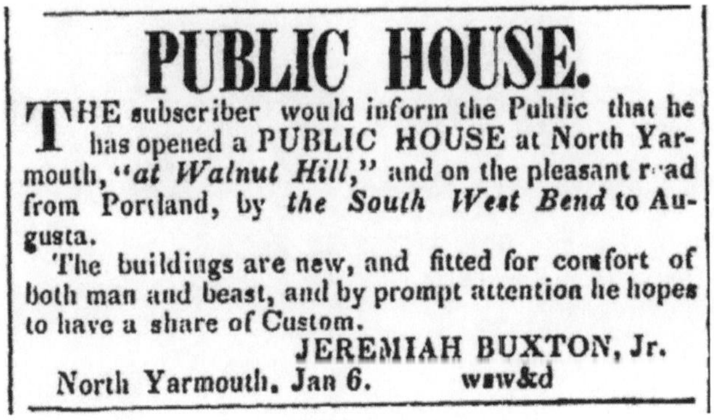

PUBLIC HOUSE.

THE subscriber would inform the Public that he has opened a PUBLIC HOUSE at North Yarmouth, *"at Walnut Hill,"* and on the pleasant road from Portland, by *the South West Bend* to Augusta.

The buildings are new, and fitted for comfort of both man and beast, and by prompt attention he hopes to have a share of Custom.

JEREMIAH BUXTON, Jr.

North Yarmouth, Jan 6. wsw&d

Public House[5]

Jones Tavern in Pownal[9]

POWNAL

Jones' Tavern

Captain Cyrus Jones kept a tavern in Pownal. He was a soldier of the War of 1812, and during that war carried a load of specie for the Government, from Portland to Canada, in a wagon pulled by a four ox team. Later he was commissioned as a captain by Governor Brooks of Massachusetts in the Third Regiment of the Militia of the Commonwealth.[6] Cyrus was also Postmaster for Pownal in 1839.[7] He died in 1846, at the age of 55.[8]

Ettie Latham tells us the following in her book, *The History of Pownal*, "It is a matter of much regret that I cannot give more information regarding that famous hostelry, Jones' Tavern, that in the early part of the last century was such an important factor in the life of our town. It was built by Cyrus Jones, sometime around 1800, and was a large and commodious house, a great ornament to the town. It had an ideal location, peculiar in a way, but exceedingly picturesque. One of the stories with which my father used to delight me when a tiny girl was of how, when a small boy, he and the other lads in the neighborhood, used to perch on the

4 Katie Murphy, *The Red House: Facts, Stories and Conjecture* North Yarmouth, Historical Society.

5 Portland Daily Advertiser (Portland, Maine) February 18, 1835 p1.

6 *The Spirit of '76* Volumes 9-12.

7 *Official Register of the United States: The Post-Office Department and the Postal Service.*

8 Find a Grave The inscription reads: "Sleep on dear husband and take thy rest. God called thee home, He thought it best."

9 Courtesy of Pownal Scenic and Historical Society. Painting by Clara L. Dyer, granddaughter of Cyrus Jones.

stone wall in front of the tavern, and watch the coming and going of the gaily painted coach with four prancing horses. One of Pownal's choicest possessions, a finely preserved old pound [cattle] is situated close by the site of this old tavern which burned nearly a score of years ago."[10]

DURHAM

Merrill Hotel

In the early 1800s, Dr. Symonds Baker built a one-story house at the corner of State Routes 9 and 136. There he maintained an apothecary shop. The house was afterward enlarged and became Durham's principal hotel, kept by Jonathan C. Merrill. Following him, Joshua Miller, Samuel Miller, John Miller and Abner Merrill kept the inn. Earlier inns at Durham included Abel and William Stoddard's Tavern, Strout's Hotel and Proctor's Hotel.

Johnathan worked at the trade of cooper until about 1820, when he moved back from Portland to Durham and opened a hotel and country store. He was in the trade and lumber business the greater part of his life, and was known as an active business man, identified with all the interests of the town. He twice represented the town in the State Legislature and was on the Board of County Commissioners. He died in Durham on 5 May of 1865.[11]

JONATHAN C. MERRILL.

Jonathan C. Merrill[12]

LISBON

Cushman's Tavern

Cushman Tavern is an historic tavern, now a private residence, on Maine State Route 9 and Webster Corner Road in Sabattus and Lisbon, Maine. This Federal House is most significant for the folk-art work along the walls of its halls and stairwell by Orison Wood, an itinerant muralist and protégé of Rufus Porter, a well-known exponent of the form.

Cushman tavern is a large two and one-half story wood frame house that straddles the town lines of Lisbon and Sabattus. The main block of the house has a side gable roof and twin interior chimneys. The house was built about 1825 by a Mr. Dwindal, and was purchased two years later by Samuel

[10] Ettie Latham, *The History of the Town of Pownal.*

[11] Everett S. Stackpole, *History of Durham, Maine* (1899).

[12] Ibid.

Bachelder Tavern in Litchfield with Adjacent Meeting House[13]

Cushman. Cushman and his descendants operated a tavern on the premises for many years. It has long since been converted into a single-family residence. The house's principle feature of significance lies in a series of murals that adorn the hallways of the first and second floor of the main block, and the stairwell that connects them. They were executed by Orison Wood, an itinerant fresco artist, known to be active in the area in the second half of the 1820s. Cushman family lore recounts that the painter arrived in 1826, offering his services. Wood was trained by his father in the art of fresco, but took his idea about content and form from the writings of Rufus Porter, whose art work still graces walls across Northern New England. These works may have been executed under Porter's supervision.[14]

[13] Postcard with incorrect location as West Gardiner. Information courtesy West Gardiner Historical Society.

[14] Wikipedia, *Cushman Tavern*. Retrieved August 1, 2017, from Wikipedia.org website: <https://en.wikipedia.org/wiki/Cushman_Tavern>.

LITCHFIELD

Bachelder's Tavern

Asa Bachelder came to Litchfield from Loudon New Hampshire in 1812, and engaged in farming and running the hotel known as Bachelder's Tavern. The tavern was built of bricks manufactured on the place and became a stagecoach stop on the Portland to Augusta route. This is an excellent example of how taverns were frequently built adjacent to the meeting house. Bachelder became prominent in town affairs, becoming the first Postmaster at South Litchfield and a member of the State Legislature.[15]

Asa died in 1846, and his tavern stand and farm were offered for sale in April of 1847. The ad carried a description as follows: "Said farm

[15] *History of Litchfield 1895 and an Account of Its Centennial Celebration 1895.*

Hallowell House[16]

consist of about 70 acres. The house is a large brick house and cost about $3000. It is pleasantly situated near the center of the town, and has good and convenient out buildings. It is on the old post road from Hallowell to Brunswick and is now occupied as a tavern and is a good stand for that purpose."[17]

Less than two months before his death, a petition was presented to the House of Representatives requesting the suppression of the sale of intoxicating drinks. The presenter was Benjamin F. Grant of Litchfield.[18]

HALLOWELL

Hallowell House

In 1814, Moses Palmer announced that he would be taking over the house lately occupied by Major J. Kidder at the sign of the Golden Ball, and would be improving it as a tavern.[19] Palmer ran it for about eight years as a tavern and stage house handing over the reins to Leonard Richmond in 1822.[20] Richmond continued the tradition for another ten years as the Hallowell Hotel, but in October of 1827 sold off the livery stable adjacent to the house. This may not have created a significant problem as the Hallowell

[16] Courtesy of Penobscot Maritime Museum.

[17] Maine Cultivator and Hallowell Gazette (Hallowell, Maine) July 24, 1847.

[18] Gospel Banner (Augusta, Maine) June 13, 1846.

[19] American Advocate (Hallowell, Maine) February 19, 1814.

[20] Hallowell Gazette (Hallowell, Maine) January 4, 1815 and January 29, 1823.

Livery stable was located just down Second Street, near the Post Office.

By 1833, a bill was introduced to the Maine Legislature to form a corporation for the so called Hallowell Hotel. Construction of a new brick building began in 1832 and was completed in 1834. It was five stories high, stretching eighty-four feet along Second Street, with a depth of one-hundred twenty-three feet including a two-story wing. The hotel contained 70 rooms and was the largest building a traveler would see between Portland and Hallowell.

Its major purpose was to accommodate members of the Legislature and other distinguished people, including Phillip Brooks, Cornelius Vanderbilt, Daniel Webster, Franklin Pierce, and Ralph Waldo Emerson. The hotel was designed by Hallowell architect John D. Lord, who had just completed supervising construction for the State Capitol in Augusta. The building became the offices of the Maine State Public Utilities Commission in 2009.[21]

Even though the Town of Hallowell did not get the first bridge across the Kennebec, or the State Capitol, they could brag that many of the Legislatures stayed in their remarkable hotel when in session at Augusta. Even Captain Palmer lived to see what his early work had yielded before he died in 1835 at the age of 62.[22]

Emma Huntington Nason, in her book *Old Hallowell on the Kennebec*, discusses the stagecoaches there.

> Sixteen stage routes centered at Hallowell, and long lines of vehicles of various kinds might daily be seen coming down Winthrop Hill or along the other thoroughfares. The main street of the town was often so crowded with these country teams that it was difficult for a carriage to find a passageway. Many of the country merchants drove directly to the wharves where the cargoes of groceries, rum, molasses, and other luxuries of life were sold before they were unladen; for all this time Hallowell was the business center for a region of sixty miles around, and from all the settlements east, west, north, and south, came the small traders and people in general to purchase their stocks of goods and all their household supplies.

She continues:

> On this particular trip, I stood one afternoon awaiting the arrival of the stage from Augusta. It was an event – it came. A bank was located nearby, and just as the stage was leaving, Mr. Artemis Leonard, its president, came briskly out, lifted the blanket side, which was hiding from view all the passengers, and said, 'anyone here going to Boston.' Someone must have said 'yes,' possibly someone he knew from Augusta, for he continued, 'I have a package here of $10,000, I would like to have it taken to the Suffolk Bank.' It was taken on and the stage departed.

Concerning the Winslow Tavern:

> About fifty years ago, a Fourth of July excursion to Hunnewell's Point took

[21] Maine Memory Network Hallowell Hotel.

[22] Daily Commercial Advertiser (Bangor, Maine) August 24, 1835.

from the town all of the fireman and indeed most of the male population. Late in the afternoon a fire was discovered in the Exchange Hotel commonly called Winslow's Tavern on Water Street. It was a large wooden building, and unless the flames could be quickly extinguished it must go, and adjoining stores also, several of which had already taken fire. Tidings of the catastrophe came to the ears of the women, and a dozen or more of the younger ones, regardless of the silks and laces in which they were robed for some function, rushed to the engine house, found men enough to drag the 'Tiger' to the wharf, and one band manned the brakes in a fashion worthy of long practice and masculine muscle, while another passed the buckets of water, and the town was saved. Six of these 'firewomen' are now living and three of them reside in Hallowell.

These hostelries of Fort and Hook, like those of old New England towns, were centers of social and political life. Here the people congregated and discussed the exciting questions of the day. Local politics, the election of a representative to General Court, or the measures of the Federal government were alike subjects of absorbing interest. Pollard's Tavern was a typical hostelry of the period. Here we are told, the men of the town often poured into their cup of enjoyment too large an infusion of artificial stimulants, and the gambols of exuberant spirits were often more exciting than commendable; but this was in the days before the temperance movement had banished the wine cup, and all the dignitaries of the town and even the divines of the church sanctioned by their example the common custom of drinking, both at home and on public occasions. What then could be expected of the common people?

The most notable, although perhaps not the most worthy but a most highly-to-be-honored guest of old Hallowell in its early days was the French statesman, Tallyrand. It is stated by North in the History of Augusta that in the year 1794, Tallyrand in company with a young Frenchman came to Hallowell, and was entertained at the tavern of Billy Pitt; that Tallyrand and his young friend dined with Mrs. Colonel North and her son, and visited Mr. Charles Vaughn at the mill house near Bombahook brook. The date of the visit of Tallyrand at Hallowell is fixed by North in the year 1794, but the Boston Press said it was October 21, 1797.[23]

A couple of other taverns in Hallowell deserve mention. As a stage stop, the House of Captain Moses Palmer must be included. Palmer took over the house previously run by Major J. Kidder in 1814.[24] In 1815 it was noted that an accommodating stage to Belfast and Castine would start from there and that the stage books would be kept at that location.[25] Captain Palmer died in 1835 at the age of 62.[26]

23 Emma Huntington Nason, *Old Hallowell on the Kennebec.*

24 American Advocate (Hallowell, Maine) February 19, 1814 p1.

25 Hallowell Gazette (Hallowell, Maine) January 4, 1815 p4.

26 Daily Commercial Advertiser (Bangor, Maine) August 21, 1835 p4.

Thomas Phillip's Tavern, at the crossroads, was previously run by Thomas Parker.[27] In Hallowell, on the Winthrop Road, Sarson Butler Junior had a tavern stand.[28]

On Water Street, a tavern stand previously run by Thomas Winslow, was taken over by B. and G. W. Hodges. They offered good rooms at reasonable prices to entice members of the Legislature. They also had horses and carriages to let from their large and convenient stable.[29]

The Back Route

By 1836, a "Back Route," this time to Augusta, was advertised.[30] This route went through Westbrook, Falmouth, Gray, New-Gloucester, Minot, Danville, Lewiston, Greene, Monmouth, and Winthrop.

In reading the ad, it is important to realize that it states that the stage passed through the towns mentioned. It does not assert that the stage stopped, or even passed through the center of those towns.

According to Georgia Merrill in her book, *History of Androscoggin County, Maine*, "The road from Portland to Hallowell was opened in 1793. It was adopted as a United States Mail Route which went from Portland to Augusta by way of Greene and Monmouth. The new route could be traveled in one and one-half days while the old route through Bath had taken two days."[31] When the road opened, only horseback transportation was used. By 1807 the roads were sufficiently passable to use stagecoaches.[33]

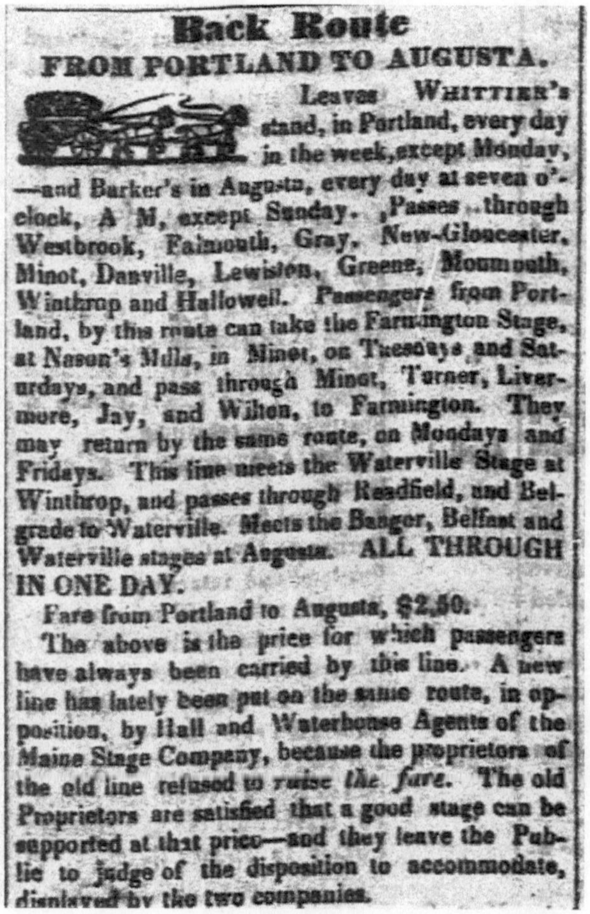

Back Route

FROM PORTLAND TO AUGUSTA.

Leaves WHITTIER's stand, in Portland, every day in the week, except Monday,—and Barker's in Augusta, every day at seven o'-clock, A M, except Sunday. Passes through Westbrook, Falmouth, Gray, New-Gloucester, Minot, Danville, Lewiston, Greene, Monmouth, Winthrop and Hallowell. Passengers from Portland, by this route can take the Farmington Stage, at Nason's Mills, in Minot, on Tuesdays and Saturdays, and pass through Minot, Turner, Livermore, Jay, and Wilton, to Farmington. They may return by the same route, on Mondays and Fridays. This line meets the Waterville Stage at Winthrop, and passes through Readfield, and Belgrade to Waterville. Meets the Bangor, Belfast and Waterville stages at Augusta. ALL THROUGH IN ONE DAY.

Fare from Portland to Augusta, $2,50.

The above is the price for which passengers have always been carried by this line. A new line has lately been put on the same route, in opposition, by Hall and Waterhouse Agents of the Maine Stage Company, because the proprietors of the old line refused to *raise the fare*. The old Proprietors are satisfied that a good stage can be supported at that price—and they leave the Public to judge of the disposition to accommodate, displayed by the two companies.

Back Route[32]

Merrill further mentions that the places where horses were changed (stage stops) included Gray, Green, and Winthrop. Although she does not include Danville in the list, she goes on to say "At Littlefield's Tavern in Danville, the stage from Farmington was met, and two coaches were generally required from there to Portland." Considering the distance from Greene to Gray, Danville must also have been a horse change stop.

Stackpole in his *History of Winthrop*, states that: "This route ran through Lewiston and Winthrop

27 American Advocate (Hallowell, Maine) June 27, 1828 p4.

28 Ibid January 28, 1826 p1.

29 Maine Cultivator and Hallowell Gazette (Hallowell, Maine) December 9, 1843 p3.

30 Eastern Argus (Portland, Maine) January 26, 1836.

31 Georgia Drew Merrill, *History of Androscoggin County, Maine*.

32 Eastern Argus (Portland, Maine) February 2, 1836 p.4.

33 Everette S. Stackpole, *History of Winthrop, Maine*.

instead of Bath or Brunswick because there were fewer rivers and large streams to be ferried."[34]

We can next examine these individual stops.

Westbrook

At the time this route was announced, Westbrook contained any route passing from Portland to Falmouth. The equivalent route on today's streets would be Congress Street to Stevens Avenue, then northeast on Allen to Washington and northwest on Lambert Street to connect on to Blackstrap Road. There were probably no stops made in Westbrook.

Falmouth

For this "Back Route" to Augusta, travel through Falmouth proceeded from the Lambert Tavern and bridge first over Blackstrap Mountain and later on around the eastern flank of that mountain.

Lambert Tavern

In 1901, a news story came out that the Lambert Tavern in Falmouth was about to be torn down. Fortunately, there were still people living who had known about the tavern, and were even related to the original owner.[35] One hundred years later, Marge Devine, then President of the Falmouth Historical Society, dug into the tavern's story and published it in her quarterly newsletter.[36]

Lambert Tavern Sign[37]

The possibility that before many months the famous Lambert's Tavern which has been a landmark on the old Portland and Gray road for nearly a century and a half is to be torn down, brings it once more into prominence and makes a brief history of this remarkable career of more than ordinary interest at the present time.

There are but few residents of Portland and vicinity who have not at some time visited this old structure, or listened to the weird and romantic tales which many of the older people tell about it. Could the shades of the departed, who, when on earth, sought the hospitality of its walls, but tell their life stories over again, what scenes

[34] Everette S. Stackpole, *History of Winthrop, Maine.*

[35] Portland Sunday Telegram (Portland, Maine) June 30, 1901.

[36] By permission from Marge Devine, Falmouth History Vol. 4 No. 1.

[37] By permission from American Equities, Inc.

they could unfold before our eyes; what mysteries they could explain.

The old tavern is located in the town of Falmouth, within almost a stone's throw of both the Westbrook and Portland Lines. It stands as has been mentioned before on the old Gray Road leading up over the western slope of Blackstrap, and just beyond the covered bridge which spans the Presumpscot River at the extreme end of Lambert street. The old inn is about two miles above North Deering, and stands alone, surrounded by tall elms, at the foot of the so-called Winslow's Hill. Its situation, in itself, is enough to inspire a feeling of mystery concerning its past, and one does not doubt for an instant the assertion which has been made regarding the gruesome looking old building that if ghosts are to be found anywhere on earth this is the place to find them. The buildings probably numbered about half a dozen, but storms and old age, have now reduced that number to only two; a tavern and a large barn. The old inn is located on a slight rise at the foot of a steep hill which stretches far up behind it. It faces toward the Presumpscot River, which is but a few hundred yards away with the main road running along its easterly side. A large L is built on to its northern end, and on to this again, is the building which contained the watering trough. Next came a small stable beside this one, another located near the road. Out behind the main house, or opposite the roadside of the structure, was the woodshed, and further off another small building which is probably a place for storing butter, eggs, etc. Across the street were two more barns, small ones, which were used for storing farming implements, and later for other purposes. Another small building was also located on this side of the road, which during the middle of the last century was used for a cobbler shop. The tavern is a large rambling structure, two and one-half stories high, made entirely of hewn lumber. Immense hewn timbers run through it from end to end, and are fastened together in a manner which was intended to withstand the ravages of time, and which certainly has accomplished their purpose. The clapboards which cover the old tavern are made in the fashion which is rarely seen nowadays, not even on the oldest of structures. Every clapboard besides breaking joints is planed down at either end to a sharp point. On one, the point comes up to the outside edge and on the other it goes down to the inside. They are then put together so as to overlap at both ends making a doubly tight joint. These old clapboards have stood the storms of a century and a half, and are still good for a number of years to come. The tavern chimney in itself denotes to some extent the great age of the house. It is one of those immense brick structures taking up the greater portion of the center of the house, and projecting out from the middle of the ridge poll for many feet into the air.

Inside were originally three main rooms and an entry, in the main portion of the house, and two more rooms in the L. The arrangement is the same upstairs, and the attic is one large room, unfinished. Entering from the front door which is on the south side, one finds himself in the entry. Going to the right a large room is encountered, which was probably a sitting room. On the opposite side of the entry is another large room, the same size as the sitting apartment. The bar room was in the back of the house, and stretched completely across the building. It was entered by a small door from the roadside of the tavern, but which during recent years has been converted into a window. The rooms are low, and are long and rather narrow. They are finished in an antique style, with brass doorknobs, which are still as good, apparently, as the day they were put on. The rooms at the present time are papered but otherwise they are about the same as when the inn was built. In the large L on the rear are two small rooms, which were probably used for a kitchen and a storehouse. In the shed which adjoins this L is the watering trough, with a log dug out to hold the water. The old trough is still filled with water from the spring which has supplied the house ever since it was built. This spring is on the side of the hill about 300 feet from the house. In every room in the main part of the tavern is a huge fireplace into which a person could easily crawl and be completely hidden from view.

Outside the building are numerous immense Elms as old as the house and which cast a grateful shade over it all times of day. There are also several fruit trees near at hand on the southerly side.

The complete history of the whole tavern is not known. It is doubtful however if anyone knows more about the past of the old house than Mr. Zacchaeus O. Lambert, now living at Deering Center. To a Telegram representative, Mr. Lambert told the following about the old inn and its interesting history:

'I cannot say certainly when the tavern was built, or who built it but it was at least 150 years ago that it was erected, and I'm fairly sure that was my grandfather for whom I was named, Zacchaeus O. Lambert, who constructed it. It was he anyway, for whom the inn was named and who ran it as long as it was run as a public house I was born there 76 years ago the 13th of next October and my father was raised there before me.

It was a great resort in the olden days and everybody for miles around put their horses up there and stopped at the bar to get their drink of New England rum; for everybody drank rum in those days. The house was on the old turnpike running from Gray and of course Gray people stopped there more than those from the other towns. Almost every night the house would be crowded and many a night have those who stopped there been obliged to make their beds on the bar room floor, or anywhere else where they could find a vacant spot on account of every room being full. In those days oxen were the only animals used for heavy loads and many a yoke stopped at the old tavern to water at the old trough in the shed. Oxen were used mostly then to haul wood from upcountry to the city where it was used in the factory's, houses, etc. Coal was almost unknown in those days and if it was to be had, its price so high that no one can afford to buy it.

I remember one incident of my own life, when I was hardly more than a baby the housekeeper put me through one of the seven by nine panes of glass in the windows. She had been locked out and she put me inside so that I might open the door for her. It was a tight squeeze but there would be a tighter one now. It would be about all I could do to get through the whole window much less a hole the size of a pane of glass.

In 1833 or 1834, I cannot remember exactly which year it was, we all thought the world was coming to an end. That was the year of the great shower of stars in this vicinity. With my own eyes, I saw them coming down on one of our hillsides, not far from the house, and the site was so beautiful it almost blinded me.

I have a dress which I could show you, which is one of those I wore while at the tavern. In fact, it is the first I'd ever put on, and it is therefore almost seventy-six years old. I have several other relics also connected with my life at the old house. The old Lambert tavern sign is in the attic of the building now.

Grandfather ceased to run the place as a tavern about 65 years ago, and it has never been run as a public house since. After it stopped to be an inn, my father, Joseph H. Lambert, lived upstairs, and my grandfather down. My father had a stone shed in the rear of the house which he operated.

After my grandfather died, father moved off the old place. The reason the old house ceased to be run as a tavern was because grandfather was getting old, and father with his other business could not attend to it.

In the olden days everybody used to drink, and they would come out there for their carousals. I remember one man in particular who came out and got intoxicated. He got so filled up he was almost out of his head, and when no one was looking he took a tumbler made of heavy glass and chewed it all up biting off a little piece of glass at a time and then throwing it out. He cut his mouth all up and slashed his face and came near killing himself before he was discovered. Large parties used to come out from Portland to go gunning, leaving their horses at the tavern.

The great event in the old tavern's history, and the one for which her name will be remembered, was from the sailors of the British warship Boxer, and the American warship Enterprise came out to the tavern the day after the fight off the harbor; that memorable battle in the war of 1812, and had their carousal. American and English were all the same that day. It is doubtful if an Englishman could have told his own countryman from that of the enemy, the American, he had just been fighting. It was an historic day for the old inn, as well as for the residents of Portland.'

'At the top of Winslow's Hill,' continued Mr. Lambert, 'is another tavern, called Winslow's Tavern, but it is not near so old as grandfather's. The old covered bridge, which has always been known as Lambert's Bridge is another historical point in the city. I do not remember when the bridge was first built but it must have been long before the tavern. I remember well when the old original bridge went to pieces. A party of teamsters had just crossed it going toward Portland, when with a crash I saw the old bridge tumble, and then collapse going into 100 pieces. The center had given away and if the accident happened two minutes sooner the whole party of teamsters would have gone down with it. It was a narrow escape for them. I remember about fifty-five years ago when this happened, and the present bridge was built right away after the old one came down

The old farm of my grandfather contains about forty-two acres. It used to produce 100 tons of hay, and that was composed of two fields and two pastures. I have seen potatoes on the old farm that were immense in size, and turned out a bushel to every four plants. After father moved off the old farm Charles Knight, who married my sister Mary G. Lambert, lived there and died there. Charles was the son of Mr. Knight, a well-known local merchant in the old days. Charles carried the farm and ran a butchering establishment in connection.

The old house is now occupied by George Knight another member of the same family. He has no family of his own, and lives alone with a young man whom he has adopted.

The old tavern is now the property of Leonard E. Pride who lives across the river and it is fast going to decay.'[38]

The other Portland newspaper, added a little more information about the tavern in the form of an old legend:

Story of an Old Crime.

Mysterious Disappearance of an Aged Peddler Who Was Stopping at Lambert's Tavern.

It was at least 100 years ago that the supposed affair took place, a period when murders were perhaps less thought of then at the present time. The story as told by some of the old residents, who used to live in the vicinity years ago, but of course, not at the time of the supposed killing, tell the story as follows.

One dark and stormy night an old peddler arrived at the Lambert Tavern, travel stained and weary. He asked permission to spend the night at the inn and the hospitable innkeeper, Mr. Lambert, gave him a hearty welcome, made him as comfortable as he could in his overcrowded house, for the late arrival at the old tavern in those days stood a poor show of finding a vacant

[38] Portland Sunday Telegram (Portland, Maine) June 30, 1901.

> bed to crawl into, and was happy if he can find a lounge, or a piece of the floor that was unoccupied. The old peddler had a large lot of goods with him and it is stated that he had quite a sum of money on his person. He also had a horse and wagon, in which he carried his stock in trade, and the animal was placed in the big barn, which stood alongside the road.
>
> Sometime during the night, a fight was started in the bar room. It is not known how it began, but it was probably a drunken row, and before the people knew what was happening, the ole peddler lay mortally wounded on the floor, how injured history does not record. Death soon ended his memories and it is alleged that his body was carried out and buried on the premises and that none but those who were in the bar room at the time ever knew what had become of the old man. Those who were engaged in the affair said that the peddler had suddenly taken it into his head to keep on his way to Portland and that was the last ever heard of him.

Another version of the story:

> Another story is that the old peddler was not killed accidentally but murdered out right. It is told that two or three learning that the old man had money on his person, enticed him out of the hearing of the rest of the inmates of the tavern and binding the old man, carried him down on the riverbank, or somewhere in the vicinity, where he was murdered in cold blood. It is said that his horse stayed around the tavern for years and Mr. Lambert could never understand why the peddler had gone off and left his horse behind him.
>
> He never imagined for a moment that the peddler had been murdered and it is a question how the story ever leaked out if true, and if not true, by whom it was started. The story of the supposed murder is an old one and has been told and retold during the past 75 years and handed down from one generation to another.[39]

Although this tavern never seems to have been used as a stage stop, it certainly qualifies as an interesting example of tavern life in Maine. The route across Lambert's bridge was, for years, the best way to get to Gray, Lewiston, and beyond.

Winslow / Allen Stage Tavern

This tavern is located just below the summit of Blackstrap Mountain some 420 feet above sea level. Although just eight miles distant from the Stage House in Portland, horse changing at that point might have been necessary due to the steep climb up Blackstrap Mountain.

"The old Winslow Tavern, later known as the Allen Stage Tavern, was built around 1809. There were five fireplaces. The room to the left of the front door was used in the building's tavern days as quarters for the stagecoach drivers who would enter the building from the street side door rather than walking around to the front door. In the driver's room were a couple of beds where the stage drivers could

[39] Portland Evening Express (Portland, Maine) June 28, 1901.

Winslow / Allen Stage Tavern[40]

catch a brief rest while the horses were being changed. Over the mantle of the fireplace is a cupboard. Inside is found a narrow shelf built around the chimney. The theory is that on this shelf mugs of toddy were stored after supper to mull and keep warm until bedtime." Just as with the Bucknam Tavern, the long side of the house was at right angles to the roadway. This may have been designed so that the stages would have a courtyard to drive into, or it could have been to keep the road noise away from the resting travelers.

William Winslow, 1750-1834, and his wife Phebe ran the tavern for some years. Together they had 18 children, roughly 10 times today's average. Winslow died in 1834, but at some previous time had transferred his property to his son, Joseph Winslow who in turn sold it to Reuben Allen in 1841.[41] Reuben then sold it to William Allen two months later.[42] William shows up in that location on the 1856 map of Cumberland County. That was a few years after the time, when the Maine Stage Company began their route to Augusta over Blackstrap Mountain. William Winslow was a member of the Society of Friends, as were many of his neighbors in North Falmouth.

The question frequently comes up "Why would a Quaker own a tavern and sell alcoholic beverages?" The answer is found in the following: "Quakers were NOT forbidden to

[40] From the personal collection of Marge Devine, Falmouth.

[41] Cumberland County Deeds, Book 170 p 547.

[42] Cumberland County Deeds, Book 172 p 122.

Hall's Tavern[43]

drink alcoholic beverages which were commonly accepted in the 1600s when the Society was founded. They were however forbidden to drink to excess. A number of early Quakers are known to have owned Inns in which the sale of alcoholic beverages along with meals & rooms was a normal part of business. Quakers did however seem to be forbidden to engage in the manufacture of alcoholic beverages. A few instances are known in which members were disowned for operating stills. It was also considered contrary to discipline to offer alcohol as an inducement to others to perform some act. One case is known in which a Friend was cited for offering alcoholic beverages to his field hands, presumably to get more than a fair day's labor for a day's pay."[44]

Hall's Tavern

This tavern was built by Nicholas Hall in the first decade of the 19th century. Nicholas was born in Dover, New Hampshire, the 11th child of Quaker, Hatevil Hall. Hatevil Hall moved to Falmouth from Dover in 1753 and settled there becoming "one of the best and most thrifty farmers in the section."

In 1823, Nicholas sold his house and 100 acres of land to his son, Ozni, the 10th and youngest child, for $2000. The son, then 24 years of age, opened his house to the public but died nine years later.[45]

It was then purchased by his brother in law, Major Edward Allen for $1400, and the house continued as an inn or tavern until the 1890s when it reverted to its original status as a private residence.[46] Currently it is still open as a Bed and Breakfast Inn known as the Quaker B&B.

43 Courtesy Quaker Bed and Breakfast.

44 *Quaker Customs & Beliefs*, Jerry Richmond, Fidonet Genealogy Conference, 1995.

45 Eastern Argus (Portland, Maine) October 31, 1832 p3.

46 National Register of Historic Places Inventory PH0508501.

Washburn Tavern / Falmouth House[47]

When stage travel was diverted from going over the top of Blackstrap Hill to the New Gray Road, The Winslow Tavern/ Allen Stage Tavern was of interest only to the locals. The Hall Tavern and the Washburn Tavern then began competing for business at a point convenient for the first change of horses.

Washburn Tavern / Falmouth House

Samuel Hicks, 1783-1856, built the tavern in the early 1820s and moved to Danville after the death of his second wife, Nancy. His first wife, by whom he had eight children, was Abigail Winslow, who died in 1834. Hicks sold the Tavern to Otis Washburn in 1836 and it remained in the hands of the Washburn family for the rest of the century.[48] At some point, it became the Falmouth House, not to be confused with the hotel of that name in Portland.[49]

Gray

The Eagle Tavern was built by Daniel Heaney before 1800, and occupied by him until 1804. Titus Olcott Brown took over in 1811 and stayed there until 1833, before conveying it to Theophilus Stimson. In later years, it became another Elm House.[50]

47 National Registry of Historic Places Inventory PH0365203.

48 Ibid.

49 Maine Farmer (Augusta, Maine) April 25, 1860.

50 Mitchell, Daggett, Weston and Reed, *The Gray and New Gloucester Register.*

Brown's Tavern and Later Elm House in Gray[51]

Being located about nine miles from the Winslow/Allen Stage Tavern in Falmouth, Gray was a good place to change horses before continuing on to New Gloucester. There were other taverns in Gray which could also have served that purpose.

Libby's Tavern in Gray was in place at the time the" Back Route" was started.[52] Cobb's Hall had been in place since before 1828.[53] Cobb's was more of a meeting place than tavern.

Earlier, in 1827, Brown's Tavern in Gray was where the Democratic Republicans of Cumberland County met to nominate Enoch Lincoln for Governor of Maine.[54] At Cobb's, the same group had also met to support Andrew Jackson for President a year earlier.

Brown's tavern was famous among the early stagecoach inns of Maine. Titus knew many people along the routes out of Gray, and the tavern became very popular. As a tavern operator, he did not appreciate people who brought food or beverages into his place. He may have lost some business due to his strict rules about this, but overall, he was very successful for his twenty-two years in business.

About 1833, he moved to Norway to live near his son, Titus Jr., Titus' other son John Bundy Brown, was the originator of J. B. Brown and Sons of Portland, still active after 189 years in business.[55]

51 1910 Postcard by G. W. Morris, Portland, Maine.

52 Portland Daily Advertiser (Portland, Maine) August 22, 1834 p2.

53 Eastern Argus (Portland, Maine) October 17, 1828 p3.

54 Ibid August3, 1827 p2.

55 Biographical Notes: Susannah Johnson Bundy and Titus Olcott Brown. Retrieved August 1, 2017, from Glenniew.Tripod.com website: <http://glenniew.tripod.com/sjb.htm#>.

Leaving Gray, the route must have followed, very nearly, present-day US 202 which currently extends to New Gloucester, Lewiston, Greene, Monmouth, and Winthrop before reaching Augusta.

New Gloucester

Taverns in New Gloucester were relatively scarce, but the first probably belonged to Peleg Chandler, who opened in 1782. At the time, courts were held in the old schoolhouse while Chandler's Bell Tavern was used for jury rooms.[56]

Nathaniel Jenks announced the opening of a tavern there in 1797. He notes that it was located on the "great road from Portland to Hallowell." While not mentioning stage travel along that road (as there was none at the time) he does state that he "has convenient accommodations for travelers and the best of stabling for their horses."[57]

By 1828, Dexter Bearce, Esq. had a tavern-house in New Gloucester, where the inhabitants of the town of New Gloucester were invited to celebrate the 4th of July.[58]

While new Gloucester was not a regular stop along this route, there were probably many interesting stories about travelers who were waylaid due to excessive snow depth on the route.

[56] Mitchell, Daggett, Weston and Reed, *The Gray and New Gloucester Register.*

[57] Eastern Herald (Portland, Maine) October 21, 1797 p3.

[58] Eastern Argus (Portland, Maine) June 29, 1827 p3.

Danville

Travelers by stagecoach were likely to find plenty of companionship at Littlefield's Tavern. It was a good stopping-off place for travelers by ox-cart, or other vehicles less spectacular than the stagecoach. Drovers on their way to Brighton with herds of cattle and sheep found Littlefield's a convenient place to spend the night, the drovers indoors and the hundreds of livestock in an enclosure outside. The two granite hitching-posts, one on either side of the front door, were for the use of local farmers who were in the habit of resting their horses a bit while they, themselves, just stepped into the bar for rum.

The old iron sign in front of the tavern gave the year as 1838, and the Owner as Enoch Littlefield, locally known as Squire Littlefield. The good Squire was always on the front porch as the stages arrived and anyone could tell from his girth, he weighed 300 pounds, that the food inside was a least plentiful.

The kitchen was thirty-five feet long, paved with brick and contained a brick oven where baked beans, brown

Littlefield Tavern-Danville

> bread, juicy meats, and turkey were fashioned.
>
> This was the meeting place for the stages to Farmington, Augusta and Buckfield. If they all arrived at once, the party was just that much the merrier![59]
>
> Rum flowed freely in the heyday of Littlefield's and no stigma was attached to the names of the men who drank it. Omar Merrow of Littlefield's Corner retells a story of his grandfather, one of the boys of the neighborhood who went along to help carry the three 100-foot chains and the jugs of rum when the distance (30 miles) was being measured from the front door of Littlefield's Tavern to the front door of the Preble House in Portland. The jugs grew speedily lighter. The five gallons of rum provided at Littlefield's Tavern were all gone by the time the surveyors reached Gray Corner![60]

Today, the only reminder is that it was located at Littlefield's Corner, on Hotel Road. From there it was only a short stroll to the Little Androscoggin River.

Greene

This was the home of the original proprietor of the "Back Route," Thomas Longley. Longley came to Greene from Waterford around 1809. There he kept a tavern which became a "jovial center around a generous fireplace." To service his stage line, he kept a large stall of horses. Thomas' son, Benjamin, continued his father's stage route for a few years before going West.

Thomas kept tavern about thirty years. Three changes of horses were made on the route: at Gray, Greene, and at the old Cross Tavern in Winthrop.

Thomas died in 1859, at Greene. His son, Benjamin, died in New Orleans. No doubt, Longley's in Green was the stage stop at that location for the "Back Route" stage line.

Captain Jacob Royal Parker, brother of Longley's wife, Esther Parker, ran a hotel at Greene Corner known as the Washingtonian Hotel. It was a large and roomy building situated on the stage highway connecting Boston, Portland, and Augusta.[61]

Winthrop

The first inn holder in Winthrop was Squire Bishop. Town meetings were held there for several years. His son, Nathaniel Bishop followed him at the tavern. He also ran a store there containing a wide variety of goods.[62]

In Winthrop, we have another example of a sturdy lady taking over as innkeeper. The Winthrop Hotel, located on the main road from Portland to Kennebec, was equipped with a large and commodious stable to accommodate Ladies and Gentlemen traveling that route. The advertisement was signed-Humble servant, E. Miller.

59 Undated clippings from the Lewiston Journal courtesy of The Androscoggin Historical Society.

60 Mary Louise Stetson, Stagecoaching in Western Maine Offered Travel Thrills for All, *Lewiston Journal* Magazine Section October 21, 1944.

61 Walter Lindley Mower, *Sesquicentenial History of the Town of Greene, Androscoggin County, Maine.*

62 David Thurston, *A Brief History of Winthrop.*

In 1842, as the Temperance Movement was gaining steam, we learn that: "The sign of the 'Winthrop House' which for so many years has been the grand Arsenal of Alcohol in this town, came down, and is now undergoing the purifying process in order to be hoisted again, not as a lure to the benighted and unhappy drunkard, but as a sign of the Washingtonian Home. A house where drunken orgies will no longer be celebrated, but a resting place where the weary traveler can find the repose he needs, and the kind care and attention of a Washingtonian Brother."[63]

Possibly the most notable hotel keeper was Deacon Daniel Carr who came to Winthrop in 1811 as a hatter. In 1819, he bought the hotel owned previously by Mrs. Flint and continued to operate it as such. After joining the Congregational church in 1832 and being elected Deacon in 1835, his hotel became the first strictly Temperance Hotel in the state.[64]

This may be the tavern mentioned before by the name of Cross' as there seems to have not been any tavern by that name in Winthrop. On the other hand, it is not likely that a regular stage stop would be without alcoholic beverages.

Determining which tavern/inn was used as a stage stop presents a very difficult problem. As roads were added and routes altered, taverns were added or eliminated from a particular stage line's stopping places. Thus, a tavern might at one time have been a stage stop, and later not used as such.

Interestingly, town historians in their writings attributed the stage stop claim to many taverns. Proving the validity of their assertions would probably need careful examination of tax records to show how many horses belonged to that tavern at any given time. On the other hand, the stage companies probably owned their horses and just boarded them at the various stops. If we have missed any important taverns in this writeup, it is unintentional.

[63] Maine Farmer and Mechanic's Advocate (Winthrop, Maine) February 19, 1842 p3.

[64] Everette S. Stackpole, *History of Winthrop, Maine.*

Chapter Eleven

Editorials and Personal Accounts

The following short items are an anthology of experiences and observations concerning travel by stage coach.

Stage Accidents and Safety

On Friday night, last the Mail Stage accidentally upset about 4 miles from this town, and with considerable difficulty the passengers were rescued from the extreme danger that awaited them– a young lady of this town was the only person materially injured; her wound we are happy to understand is not dangerous. In consequence of the falling of the mail bag into the water remaining by the side of the road, all the newspapers destined for the Eastern part of the district were destroyed–the letters sustained no injury. We understand measures have been taken to prevent a similar occurrence.[1]

We understand that on Wednesday last, the horses in the Eastern Stage in descending the hill upon the north side of the Presumpscot Bridge, set out upon the run from the top of the hill, at which time the driver lost control of them, and thus situated, they ran furiously down upon the bridge, the leading horses running on each side of the brace of the middle post to the bridge, against which the carriage, with five passengers in it, was dashed to pieces and a Mr. Wellington of Fairfax, one of the passengers, had his thigh broken, and himself and other passengers were considerably bruised. What is surprising, but one of the horses were so disabled as to prevent their proceeding.[2]

The horses of the Eastern Stage, took fright yesterday, in coming from Gardiner, ran away with the carriage, and threw the driver and passenger from the outward seat, both of whom were slightly injured. One or two passengers who jumped from the carriage, were considerably bruised, while those who remained within, as most commonly happens, escaped unhurt.[3]

1 Eastern Argus (Portland, Maine) September 21, 1816 p3.

2 Hallowell Gazette (Hallowell, Maine) September 24, 1817 p2.

3 Portland Advertiser (Portland, Maine) October 23, 1824 p2.

To the Honorable Legislature of Maine.

Among other subjects worthy of your consideration, permit me to mention the subject of our stagecoaches. It has become so much the fashion for people to make use of this mode of conveyance, that stagecoaches are multiplied on all our principal roads, and the proprietors seem to meet with ample encouragement. The increasing number of these vehicles is proof that they are profitable: I am glad they are so. But they ought also to be safe, convenient and comfortable; and certainly, in many instances they are not so. The time seems to have arrived when it has become necessary for the Legislature to interfere and establish some regulations which are important to the community and especially to all those who availed themselves of the opportunity of traveling in stagecoaches. Before I would venture to recommend any particular regulations on this subject, I will mention some of the existing evils.

In the first place, steady and accommodating drivers should always be employed. And although in this part of the country, such are generally employed; yet this is not always the case, they are not always careful, even if they are honest and steady. On this head however, there is little ground for complaint. But in the second place it is a notorious fact, that our accommodating stages, the greater part of the time, furnish little else than inconveniences to the traveler. Passengers are crowded in, in such numbers, that they are suffering a species of imprisonment. The fact is, the profits of the owners only are regarded, and the comforts of the passengers wholly neglected in making the necessary stowage. Such conduct as this is an imposition upon the traveler to which he ought not to be subjected; and it is one which he is often compelled to endure, after having paid his fare. One stage proprietor adopts this course because another does; and this competition only increases the evils and dangers. Often as a stagecoach comes into this town, with ten or twelve passengers crowded into the carriage in a most unmerciful manner and several sitting with the driver or seated on the top, thus exposing all concerned to the dangers consequences on breaking any part of the carriage.

But, in the third place, this practice is highly reprehensible on another ground. All our laws consider it an offense to treat beast with cruelty. The laws of humanity also forbid it. And yet how often do we see four stage horses dragging a heavy carriage, with fifteen or sixteen passengers, over roads often hilly and muddy, and almost worn out with fatigue, and in the heat of the summer, dropping down and dying in the harness, and the faithful and excessive service of an avaricious and unfeeling proprietor. These things ought not to be so; and it is for your Honors to correct these abuses, and place our stage coaches on a respectable, safe, comfortable and humane foundation.

And now permit me respectfully to suggest to your Honors a corrective of the abuses above mentioned.

Let a law be passed providing:

One. That no person shall be allowed to drive a stage, unless approved in writing by the selectmen of the town to which he belongs as a person of good moral character, of steady habits, and a safe and careful driver.

Two. That the proprietors of all stages shall receive licenses therefore, free of expense, from such person or persons as may be designated for that purpose; which licenses shall express the number of passengers that each carriage so licensed shall carry in the body of the carriage.

Three. That only one passenger shall be permitted to sit with the driver on his seat, and none on the top of the carriage, so that he may always have the command of the horses.

A law bracing these regulations, I am satisfied would be salutary in its consequences to all concerned; nor do I believe it would be objectionable. Sure I am, that I have often heard extensive proprietors express a strong wish for such an arrangement, so that existing evils might be remedied, and a fair competition be seen, without producing so many inconveniences and dangers to passengers, and cruelty and sufferings to the poor horses employed in such severe service.[4]

[4] Portland Advertiser (Portland, Maine) December 30, 1825 p2.

Stage Coach Case Law

Important to stage owners. An interesting case was on Monday decided in the Supreme Court of the State, sitting in this city (Boston). It was a prosecution against the proprietors of the Boston and Providence Citizens' Line. The plaintiff took passage in one of their coaches, and on Boston Neck, was overturned, thrown under the carriage, had his legs severely fractured and was otherwise injured. An action for damages resulted yesterday, in a verdict for the plaintiff of $1500.

The council for the defendants argued that the overturning of the coach was an accident, which no care could have avoided, and the injury was owing to the carelessness of the plaintiff.

The judge in his charge very clearly pointed out the responsibilities of the proprietors of stages. They are bound to prepare proper accommodations for passengers; to have good and sufficient coaches, horses, harness and drivers, to keep them always in such repair that full dependence can properly be placed upon them, and in case of the overturning of the coach, running away of the horses, breaking of the harness, or breaking down of the carriage, an injury thereby incurring to passengers, the proprietors are liable, unless they can prove the accident to have arisen from causes wholly without the control of the driver. But it is not sufficient that the agents of the proprietors are properly directed; They must carry their orders into execution. If the carriage breaks down while upon a smooth road, the presumption in law (as laid down by the judge) is, that it arose from neglect on the part of the owner of the stage, and the burden of proof lies with the stage proprietors to rebut the charge.

The result of this case is of great importance to the public, whether interested in stages as proprietors or passengers. Decisions have been made in England which have established the principle of responsibilities of stage and steam boat owners; but we have now for the first time a decisive opinion upon this subject, and learn the law in this Commonwealth to be the same in this respect, as that of England. Accidents in steamboats and stages have become too frequent, and it is full time that the public are informed that the law will afford them some relief against the carelessness of stage drivers and steam boat engineers. Loss of limb, or a mame for life, it is true, cannot be compensated by pecuniary restitution; but as great expense and loss of time are always attendant upon extensive injuries, it is some satisfaction to know that a remedy in law lies against the employer of reckless drivers.[5]

Fire! Fire!!

On Thursday evening the driver of the Eastern Mail, when about four miles from the city of Portland, discovered an alarming fire in his stage. There were five or six passengers all asleep and it appears that the fire had communicated to the straw from a foot stove use by a lady passenger, and was blazing violently when discovered by the driver. After arousing and extricating the passengers, he was obliged to throw snow with great activity, to save the stage from destruction; as it was, no injury was sustained, except the loss of curtains, lining etc. of the stage and the whiskers and eyebrows of the passengers.[6]

More Stage Accidents.

We understand that the same stage, but with another driver, which upset between this city and Gray on Monday last, upset again yesterday, by the breaking of the poll on the return between Gray and Hallowell, and one of the passengers, a lady, was considerably injured. We also learn that the opposition stage driven by Longley on the same route, was upset yesterday between Hallowell and Portland but the passengers came off uninjured.[7]

Safety

The apparatus for checking carriage wheels, invented by R. Jarvis, Esq. is worthy of public attention. It can be fitted to any vehicle, old or new, for a moderate expense and is easily managed.

It will be useful in confining horses, or preventing them from running away when harnessed to vehicles and left standing without being tied. In a city,

5 Bangor Weekly Register (Bangor, Maine) February 22, 1831 p2.

6 Eastern Argus (Portland, Maine) January 21, 1833 p3.

7 Ibid October 4, 1833.

horses left untied are continually exposed to fright, and in running violently through crowded streets, endangering the lives of all near. In such cases, almost the only chance in safety is in flight. By this apparatus, both wheels of the chaise, cart, our other two wheeled carriage, and the hind wheels of coaches, barouches, wagons and other four wheeled carriages, are checked and prevented from revolving, which entirely prevents the horse or horses from backing, and prevents them from advancing faster than in a slow walk, in which they can be stopped without danger.

With this invention, one may leave his vehicle in a crowded street, without any apprehension of injury to his property. This apparatus will be useful to physicians and others, who, having frequent occasion to leave their horses, and are obliged to confine them by a weight, which besides being burdensome to the vehicle, affords but little security. It will be useful in confining horses when frightened and ready to run, or when vicious and preparing to run, or when the reins, bridle, or any part of the harness or vehicle suddenly breaks. In all such cases the wheels being checked, the horses are confined until quieted or until any disorder with the harness, or a part of the vehicle is repaired.

It will be useful in stopping horses when actually running. The wheels being checked, the horse or horses can barely drag the carriage at a slow walk, and can then be easily commanded.

It will be useful in descending hills with heavily laden coaches or wagons, when locking or chaining wheels is necessary. Both hind wheels can be locked by the driver without leaving his seat. It can also be applied to the fore wheels in a manner that will enable him to lock one or both at pleasure.

Applied to stage coaches, it will put the horses under the command of both passengers and driver; the means for checking either forward or hind wheels terminating inside the coach, within reach of the former and also forward within reach of the latter. Should the driver, in stopping, leave his horses untied, or should he attempt to race, or be thrown from his seat, or should the horses attempt to back over a bank or a precipice, the wheels can be instantly checked by the passengers. So, should the coach contain no passengers and the horses should attempt to back or run, the wheels can be checked by the driver.

It will be useful on fire engines, in suddenly stopping them when necessary, and thus preventing the accident that sometimes occur from their rapid movements.[8]

Stage Accident.

We learn by a gentleman from Maine that the Eastern Mail Stage, on its way to Portland from Bangor, on Monday evening at ten o'clock, and when

[8] Jeffersonian (Portland, Maine) May 26, 1834 p3.

within a mile or two of North Yarmouth, was thrown down twenty-five or thirty feet deep, and what is very remarkable, not one of the passengers or horses was injured. The accident occurred during a severe storm of thunder and lightning. It was extremely dark at the time, and the driver found it difficult to keep the road, especially as his horses were much frightened. Finally, the horses suddenly started at a flash of lightning, and three of them jumped over a fence carrying with them down a steep bank the stage, driver, and four passengers, but all escaped injury. One horse broke from the harness, and remained in the road. The driver was held on the stage by the boot and the passengers were all inside. Our informant was one of the party, and thinks that he and his associates have had a most miraculous escape, particularly as the stage fell upon rocky soil. The accident happened within a mile of one of the regular stopping places, and in the course of an hour or two the passengers were again on the road to Portland.[9]

A Letter Home

In an 1842, letter to his wife, Charles Northend Cogswell commented that his stagecoach had tipped over near Green. He was traveling from South Berwick to Augusta as a Representative to the State Legislature. Apparently, he was a man not easily disturbed by adversity. He states that "Once the stage tipped, two large men fell on me with no major injury to myself. It only made me see stars for a minute or two but no skin was broken. The other passengers were injured a little bit but nothing to hurt."[10]

News from Augusta.

The Bangor stage, in leaving town last Tuesday, broke down on the east side of the river [Kennebec River] and killed one horse, and broke one of the ribs of the driver, Mr. Holmes.[11]

Letters Written by Travelers on Stages

Extract from a second letter received from a Bostonian now on a tour through the State of Maine, to a friend in this city.

I have just arrived in this town [Brunswick,] and took my proposed walk to Topsham. This is a smaller village then Brunswick, but very handsomely situated, the main street running parallel with the beautiful river Androscoggin, on each side of which are many stores, mechanic shops, lawyers and doctor's offices, etc. The Congregational Society have within a year or two erected a very large and elegant house for public worship, and have settled as their pastor the Rev. Mr. Costs, a very worthy, good man. The courthouse, meeting house, and many other public buildings are situated on a beautiful

9 Eastern Argus (Portland, Maine) July 18, 1835 p2.

10 Cogswell Letters, Maine Historical Society.

11 Gospel Banner (Augusta, Maine) December 4, 1847.

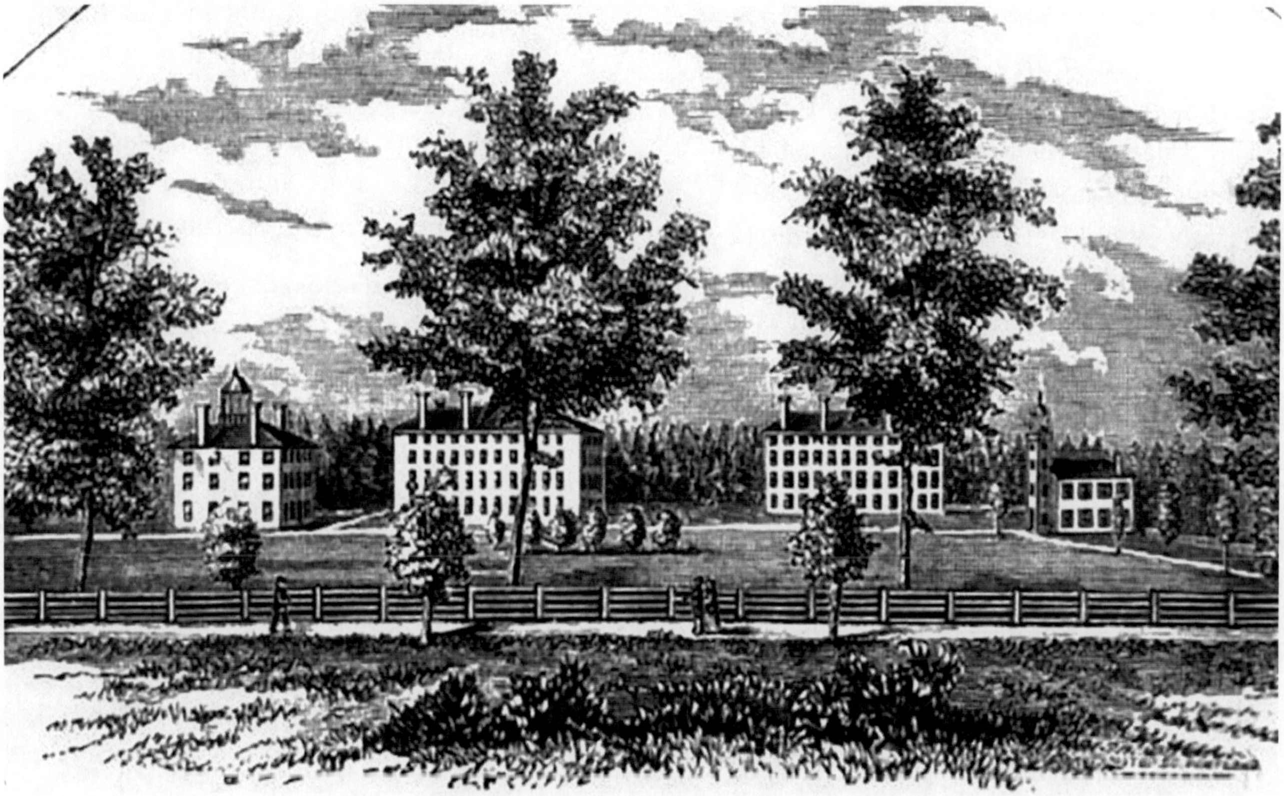

Bowdoin College 1821—Brunswick[12]

site but a few rods from the river. After a two hour walk through the town, I returned to my lodging's in Brunswick, [Mr. Moorhead's] and justice to my own feelings induced me to say, this is one of the best public houses I've slept at since I left Boston. I remained in Brunswick until Monday morning when I took the mail stage for this town [Augusta]. There are ten stages that arrive and depart from Brunswick daily. We passed through Topsham and Bowdoinham. In the latter town is the residence of the honorable Ebenezer Harrick, a member of Congress. It is not a large village, but a very handsome one; many stores mechanic shops, etc. We continued on through Bowdoin, Richmond, Litchfield to Gardner, this is a delightful town, they have many churches, a bank and a very excellent and useful institution, the Lyceum, under the charge of the Rev. Mr. Hale. We then proceeded to Hallowell where I spent the night, and this morning came to this town. I have at present only time to say that the beauty and splendor of these three towns, Gardiner, Hallowell, and Augusta, far surpass anything I ever saw before. I think I shall remain here a day or two, and then proceed to Waterville, and take an easterly route to Bangor, etc.[13]

[12] George Augustus Wheeler, *History of Brunswick, Topsham, and Harpswell, Maine.*

[13] Columbian Centinel (Boston, Massachusetts) August 6, 1825 p2.

Extract from a letter written on a Journey East

I started Friday at 10:00 AM from Portland in the accommodation stage, which is more properly so called than any other establishment of the kind in the state, both as respects the horses and carriage, and the driver is inferior to none in New England in carefulness of driving and care of his horses, which is the best trait of a stage establishment. The horses being kept well will perform their route with ease and punctuality.

After passing through the pleasant towns of Falmouth, North Yarmouth and Freeport, over good roads, we arrived at Brunswick, twenty-six miles from Portland at 2:00 PM, which is pleasantly situated on a sandy plane. There are two large buildings of four story and three story buildings for colleges, which are occupied by the students, for libraries; the old college is principally occupied with chemical apparatus, and galleries for painting and minerals of most every description. The college buildings are very commodious and pleasantly situated on rising ground at the entrance of the town. East of the college grounds is a beautiful grove of forest and other trees which has quite a romantic appearance, and is a pleasant walk for the students as a recreation from their studies. The town is principally built, on two long and broad streets, which as the soil consist of sand, are very easily kept in repair. Topsham is situated about a mile from the colleges connected to it by a long bridge over the Androscoggin near which at the falls are mills of various kinds and a large cotton and woolen factory in successful operation.

Having stayed one night at Brunswick, started in the mail stage for Bath, the horses and carriage where much inferior to those of the accommodation line. On entering Bath from the west, the town has a beautiful appearance, the meeting houses are on much higher ground than the town, from which there is a fine view of the settlement, the houses are built quite compact and on narrow streets and surrounded by hills in a low situation. The ferry across the Kennebec is about two miles from the village which made the ride about two hours from Brunswick. Crossing the ferry is rather dangerous but as the men employed are acquainted with the eddies and tide, there is not much to be feared on a pleasant calm day; the stage was in waiting at Woolwich which started immediately after landing from the ferry boat. I could not have time to look about but apparently Woolwich is a pleasant place in the summer, and rather cold and exposed in the winter; my ride to Wiscasset was over good roads, but I cannot say so much in favor of the conveyance; however, I arrived safe at Wiscasset at 2:30 having been two hours traveling seven miles; there is not much business done here at present, except in building vessels. There is a new and beautiful courthouse, two meeting houses, a large town hall pleasantly situated, an academy and three commodious public houses for entertainment.

At Wiscasset, some of the best vessels are built, both in model and workmanship that sail out of any of our ports. It is situated about 10 miles from the ocean, the river is safe as a harbor, and ships of the largest class can come up to the town with great safety. The fort at the entrance of the bay with little care and expense might be preserved from ruin and add much to the scenery about the place.[14]

Short Tour to Maine.

We've reached Bath at 5:00 AM, just 24 hours after we left Boston, including our detentions at Portland and other places. At Bath we anticipated the pleasure of joining a party of ladies and gentlemen in an excursion up the river to Gardner, Hallowell and Augusta, and there meeting on board the steamboat, the taste, beauty, fashion and intellect of the Kennebec; But in this we were disappointed, the expedition having failed at Bath by some cause or other we did not investigate. At this disappointment, our determination was at once made to take the stage for Portland, but as six or seven hours must elapse before the stage would pass through the town, we called on some friends and took a look at the place.

Bath is a pleasant town, on the right bank of the river, containing about 4000 inhabitants, and appears to be in a flourishing condition. Its growth however, has not been uniformly prosperous. The calamities of the late war reached this as well as other places. Bath, however, must grow as it has a large back country in a flourishing condition. The whole of Maine suffered for several years from what was called the 'Ohio Fever,' which was an infatuation for emigrating to the westward. The western country had been represented to them in the most glowing colors and nothing but plenty and happiness were to be found there. So many of the deluded adventurers met misery and death on the road, that after a while the disease abated. It required many a doleful tale however, from those who had tried the experiment, to teach the remaining enthusiasts the truth. The infatuation was a sad one. In the year 1807, in one day, we saw seventeen families crossing the Alleghanies. Most of them had come from Maine and a few from New Hampshire and Connecticut. It was a painful sight to witness. Many families who were well situated in a land of morals, laws, schools and churches, wandering for what they knew not, having no settled plan or distinct calculations in their mind. The evil is now gone away and is only to be remembered as a beacon for the present generation. The soil of Maine may be made as profitable to the agriculturalist as any in the west, if properly cultivated; and the state of society here is better than it will generally be in the Western Country for half a century to come.

[14] Portland Advertiser (Portland, Maine) October 4, 1825 p1.

Having traveled to Brunswick we next caught the stage for Portland. From Brunswick to Portland we passed several places, which, probably might have afforded some materials for observation; but our mode of conveyance would not permit us to gather them. The company in the stage, however, was agreeable, and in no small degree intellectual. Of the five, three of the passengers were of the profession of medicine and were conversable and instructive. There has been a great improvement within the last fifteen years in stagecoach conversation, go in any direction you will, which marks very decidedly, in our opinion, the progress of refinement and intelligence. The common topics now, would have been considered as pedantic a short time since, as being above the ordinary comprehension of travelers. Men were flying along on business and thought of nothing but overcoming the greatest portion of space in the least possible time; but so many now travel for pleasure and information, that the businessman takes the same air of one of leisure, or one in pursuit of knowledge; and in fact, is nearly identified with him in his attention in whatever passes, or in inquiries of whatever is worth recording.

It was a warm afternoon, when the shadows had become considerably lengthened, that we first saw Portland at a distance. Its first appearance is pleasant from this view, but not imposing. This evening we wandered awhile about the town but prepared the next morning to take a more distant view of it. Portland may consider herself a part of the holocaust which was offered to propitiate the fates in the revolutionary struggle. The burning of Charlestown and Falmouth fixed the determination in the timid and wavering in our country, when other arguments had failed. These wanton acts kindled a flame which at this hour has scarcely ceased to burn.

The style of building here has much improved, as elsewhere, within a few years, and one finds convenience, neatness, taste and often splendor, in every step he takes through the town. The public buildings are highly respectable, in their general appearance, and some of the new churches are truly tasteful. These things are well, are excellent; but after all, it is the intelligence, morals and spirit of the people which constitute a town.

In the morning, we shall take the stage for Portsmouth, reflecting how much can be seen, felt and enjoyed in a short time, when the mind is awake and the heart open to receive impressions. The longer a man lives the less he thinks of life, if he reasons rightly; but the more he sees of the world the better he thinks of mankind.[15]

Incidents of Travel Down East.

I don't know but incidents and accidents, may be gathered from travel in our own state as well as in foreign

[15] Portland Advertiser (Portland, Maine) July 18,1826 p1.

states. I have therefore set down during a leisure moment, to see what can be gathered from a short ride not a great way from home among our own people. Nothing could have been more agreeable, than the weather for the last days in January to a person who was obliged to go abroad, for while the sleighing was excellent a little distance east of your good city, the atmosphere was not so inclement as to destroy the comfort of using it. I was once, as I find children are now, so fond of a sleigh ride that it made no difference to me what kind of weather it was, so that I was behind a good horse and the merry jingling of the bells. But I have come to that condition when I fully realize the feeling of the humorous gentlemen, that I used to listen to with great pleasure when I was young; who used to say that he had rather stay at home with his feet in a tub of snow than to take a sleigh ride for pleasure.

I found at Augusta the principle public houses were crowded, so the poor traveler after a long stage journey, found but little chance of thrusting himself through the throng that was occupying not only the best, but all the places around the fire. I don't know any situation so desolate, as to arrive in the evening at a public house, a stranger and shivering with cold, and to meet there no welcome reception; to find yourself in the middle of the bar room, the observed of all observers, every eye measuring your dimensions, watching your movements and making remarks in an undertone which the new comer cannot but imagine are applicable to himself.

The various persons who command the crowd at the public houses in Augusta are gentlemen of the Legislature, who form, of course, the largest proportion of the assemblage, gentleman who are seeking office and others who are seeking for their friends; and a very small part of the crowd, I imagine is there, without some specific object; as the inducement to remain there has not sufficient power to hold one a single day beyond that in which his business may be accomplished. These persons employed their time in a variety of ways and present to a scene, which to one who looks on with perfect indifference, has not a little of the comic material in it. Some are lounging upon sofas, some reading newspapers, some hurrying back and forward with a very anxious and busy air as though they had the budget of Ways and Means to support against the deadly opposition. Others are collected in groups discussing some political measure or some new nomination, but the most attractive portion of the scene is the checker players! From the interest which is taken in the game both by the players and the lookers on, you would suppose that this is the mode which our grave legislatures had taken to settle important questions of public policy. All these movements are going on in the public rooms; what passes in the private rooms, it is not for the pen of the passing traveler to record.

Among the subjects which formed themes of conversation at the capitol was Mr. Otis's bill for electing county officers, viz., the Commissioners, the Clerks of the Courts and County Attorneys. The bill I believe requires them to be elected once in three years. If they will fix it at five, I will withdraw my opposition. There is much difference of opinion on the subject among persons of all parties, and it has the rare merit of being important, without being made a party question. The nomination for the office as designated in the bill, will be suspended for a while, from a delicacy on the part of the Governor who does not wish to interfere with the action of the legislatures in this matter.

Another important question relates to the creation of a new judge for the Supreme Court. Great differences in sentiment prevail on this point; some are sanguine that it will and ought to be adopted, others think it's very doubtful whether it is necessary or expedient. My own opinion is, that the addition will not be made; and within an able court, such as we shall probably have before the close of the year, I am inclined to believe that if dockets will be properly cleared and we shall hear no more of the law's delays which have been so vexatious and expensive for the last half dozen years.

But I have to go further east, I will take leave of the seat of government for the present, with the hope of returning to it again on some future occasion. Early in the morning, I mounted the box with the driver to enjoy the fresh morning breeze, rejoicing to breathe pure air of heaven, after inhaling for twelve hours the corrupted and tobacco impregnated atmosphere of a crammed hotel. The driver was somewhat taciturn but not morose, and it was only when the qualities of the horse were alluded to, an animal which excited all his enthusiasm, that he would hazard a remark beyond a monosyllable. 'A horse,' said he on one occasion, 'is a knowing creature.'[16]

Winter Traveling in Maine.

Started from Brunswick at 1:30, and reached Bowdoinham (eight miles) without trouble, the people of Topsham having shoveled and broke out the road with commendable spirit. For the first four miles after leaving Bowdoinham, got along very well but when we got to the Richmond line, the drifts were enormous. There was very little snow in the fields, the stubble was hardly covered, but the road was blocked to the height of the stone walls. Nothing had been done to render the road passable. The inhabitants reasoning, with some plausibility, that it was no use to try to clear the road while the snow was blowing at such a rate. Had now some music. The order was first to throw our weight on one side, and anon on the other, to prevent oversetting, which would have been a serious matter with such a quantity of luggage. At one time, we had nearly accomplished that dreaded maneuver and all passengers,

[16] Portland Advertiser (Portland, Maine) February 9, 1841 p4.

male and female, were clinging to the upper side of the stage on the outside, when the horses with a spring broke the harness, and we were obliged to hold on, the wind blowing a hurricane, till the harnesses could be repaired, and the snow shoveled from under the stage.

After unloading and repacking some twenty times, with the aid of people on the road, who came to our assistance with shovels and levers; and with the most persevering and judicious efforts of our excellent driver, we overtook about 5:00 PM, the mail stage and passengers which left Portland at 11:00 Friday night, nine hours before we started. Finding his horses used up, the other driver had detached them from the stage, and gone after oxen to help him out. He had sent the letter mail ahead in a single sleigh. Finding the stage in his way, Captain Sawin gave orders to capsize it out of the road, which was accordingly done, and we passed as it lay on its beam ends. Our triumph was short, however. After going a few rods farther, we were obliged to take to the field, and when we came into the road again we soon stuck fast, our horses were used up and done over. We procured oxen and made our way a short distance when, about 6:00 finding ourselves in the vicinity of comfortable quarters, we gave up the hope of proceeding farther, and were received and comfortably provided for in the private dwelling of Mr. Smith, about eight and one-half miles from Peacock Tavern, in Richmond. This worthy gentleman and his wife did everything in their power to make us comfortable, and well did they succeed, for they had the ability as well as the disposition to provide for us. A snug house, good provisions, served up in the neatest manner, with good beds and snow-white sheets, we had much more to be thankful for then we had a right to expect. I am sure their guests will ever retain a grateful sense of the kindness of this worthy couple.

Captain Smith was early abroad, and started all he could to break out the roads. He came with six oxen for his load, placed the baggage on an ox sled, with one yoke of oxen, and attached two yoke to our stage containing the passengers, but we soon found them insufficient, and were obliged to retain a third pair of oxen to get the stage along.

In this slow manner we've reached Peacock's tavern, where we were assured that the worst was passed and took horses. By unloading when we came to a bad place and the use of shovels, we got through to Gardner at 12:00 AM, where we found the road clear and reached Augusta before 1:00. Had I room to detail some of the dialogues between our driver and some of the people on the road they would amuse you and your readers. 'Boy' cried the driver to a towheaded urchin 'who is road agent in this district?' 'Father.' 'Well, you tell your father if he don't break out this road before I come along again I will welt his hide well; where is he now?' 'Gone to meeting,' answers tow head. 'Well,' says Jabe, 'he ought to be at meeting the rest of his days if he has not performed his other duties better than those of road agent; remember and tell him what I say.' 'Yes sir.'[17]

Sketches by the Wayside-A Journey from Boston to Hallowell by Railroad and Stage.

I will venture to ask a small space in your columns to recite the incidence of a recent journey from Boston to Hallowell. By way of exordium permit me to say that when comfortably seated in the car, and traveling at the rate of twenty-five miles per hour, my mind was forcefully carried back to that period in the history of our town, not yet very remote, when a journey to Boston and back again was regarded as a matter of public concern and, was talked over weeks in advance, and preparations for the stupendous undertaking were much more ample than are now thought of in a voyage to Europe.

On Saturday, the 30th of January in company with a noble band of Kennebeckers (and you remember that the facetious John Helms used to say that the Kennebecers were a head and shoulder taller than the people of any other county in the state,) I took my departure from Boston to Hallowell. Nothing remarkable transpired in the cars, but the company were instructed and entertained by learned exposition of the nature of gun cotton, its explosive character, and the other remarkable properties of this common compound by Mr. Hess of Boston, formerly a Kennebecker, who exhibited a surprising knowledge of its history and its efficacy. On our arrival at Portland about the usual hour, we were taken up to a public house and had the privilege of dining from off the carcass of a rooster who, one of the company (not having the fear of the landlord before his eyes) declared he had heard crow one spring morning about eight years since from the top of Mount Joy. After finishing our repast, a remark was made to this effect, 'that if our Portland friends had employed some of their loose coins in the construction of a hotel which should comport with the size and character of the city, it would have been a safer and more profitable investment than that P. & M. Railroad stock.'

At about 2:00 our gallant company took the stage and everything passed off as merry as a marriage feast until we arrived at the village of Cathance and here we had the first inkling of what was in reserve for us if we ventured further that night. But who ever heard of a Kennebecker being daunted? Fast horses were speedily attached to the stage and we were once more comfortably seated with coon skins and bear skins for our comfort and protection against the cold. On we went consoling ourselves with the reflection that when the elements attempted any obstruction to a Kennebecker they would find they

[17] Maine Cultivator and Hallowell Gazette (Hallowell, Maine) January 13, 1844.

had 'woke up the wrong passenger.' Thus seated, and thus happy, thinking of home and all its endearments we were suddenly aroused from our reveries by a cry from the outside of 'all hands to the rescue.'

Upon taking the latitude and longitude of our position, we found that our horses and the vehicle attached to them were buried in a snowdrift about eight feet deep and not 'few left of the same sort ahead.' And now commenced in earnest the siege of 'Monterey and the storming of the Bishops palace.' For an hour and one half we struggled to extricate our horses and carriage from their awkward and uncomfortable position. As for ourselves, at one moment some of us would be gazing upon the moon which shed its silver beams brightly upon our apparently hopeless condition, at another, someone of our company was out of sight, gone down it would appear in a fruitless effort to fathom the depth of a huge snow drift and tried to touch mother earth with their feet. The only evidence of civilization that presented itself and the only insurance that we had that we were in a land which had been at any period of time inhabited by human beings, was something that appeared in the distance resembling the ridge poll of a dwelling place just emerging above a huge draft of snow. For that spot we directed our footsteps, and upon reaching it raised the cry of 'Help!' 'Fire!' and as the last resort, 'Murder!' but no sign of vitality exhibited itself and no one responded to the call. The next day the old woman who inhabited the premises said she distinctly heard the outcry, but such things did not move her, 'she was used to them.' In this dilemma, the only alternative left for us was to endeavor to reach that hospitable home so well-known to all the Kennebeckers, as Mustard's Tavern.' Forward was the watchword. It fell upon my ears like a death knell, but I had a conference with my feet and invoked them by every motive I could address to their patriotism to 'push on and keep moving,' but they often failed me and prostrated my suffering body in a deep avalanche of snow. I looked after my companions in suffering and involuntarily passed a eulogium upon each of them as by turns they disappeared from my sight and went down to that country from 'whose bourne' no stage traveler, under like circumstances, ever returned. All their failings, if they had any were at this trying moment forgotten, and I can only call to mind their manly virtues and the deep grief of their friends at their untimely end. To perish in snowdrifts and have our dead bodies dug out per order of the School District Committee was an awful thought but such at one time appeared to but be our inevitable destiny. At length, however, through tribulation and anguish, through snow drifts and pelted with frozen particles of snow driven with great force by the wind, we've reached a haven of rest. A blazing fire and a cordial welcome soon made us forget our sufferings, and retired to rest, to dream of home and that happy hour when our journey should be terminated. At about 11:00 the next forenoon our hearts were made glad by the sight of our long-lost stagecoach drawn up to the door of our hotel by six yoke of oxen. A blooming girl of sweet sixteen never welcomed her lover with more true joy then did we this old acquaintance of ours. During our absence, any straggler if he chanced along might have converted the same into a Bankin' House and he would have found ample funds to carry on his business and meet any emergency in the money market. There would have been no danger of a run for species, or other run in such a locality. Our eyes were now turned once more homeward, and the hope strong in our bosoms that we should one day or another reach the banks of the Kennebec. That hope speedily ripened into fruition and with that loyalty so peculiar to the heart of the Kennebecker all exclaimed, 'There is no place like home, sweet home.'[18]

A Trip by Stage from Portland to Augusta 1834.

Our daring young adventurer, by the name of Daniel Cony, had booked passage on the ship "Nelson" to travel from Boston to Augusta. A threatening nor'easter called for a pause at Portland where Daniel opted for travel by stage to complete his trip to his home in Augusta.

[18] Maine Cultivator and Hallowell Gazette (Hallowell, Maine) January 13,1844 p1.

The sea part of my journey having terminated, nothing remained but to thank Captain Heath for the share he had contributed to the pleasure of the voyage, when I stepped into the boat, was landed at the wharf, and soon found myself at the stage house (Elm Hotel, I think,) where I booked my name for a passage to Augusta by the next morning's 'accommodation,' which was to leave at 7:00 AM.

Stagecoach traveling in general and especially upon all great thoroughfares, in any sense at all applicable to that phrase of one-half century ago, has become obsolete. On many routes, it was the only method of conveyance during the entire year, while upon others it only had a rival in the summer season; the slow-going packet. Although rather fatiguing, it was after all, the most jolly way of getting around that ever was invented. If the young folks of today imagine that their grandfathers never had any enjoyment in traveling, they labor under a slight hallucination. On a fine summer day, over a good road, nothing could be more exhilarating than a stagecoach journey. There was a sociability about it that somehow does not attach to the steam boat or railroad. A brief description of my ride from Portland to Augusta will suffice to show about how the thing was done by 'the fathers.'

An early breakfast at the hotel, prepared expressly for those who were to leave by the stage, having been partaken of, the customary announcement of 'stage ready' was vociferously made by the company's agent, as he paced back and forth on the long piazza of the hotel with an air of dignity and conscious responsibility peculiar to his class. There were eleven passengers all told, nine inside and two out. Honorable John Holmes, the politician, wag and storyteller, par excellence, of his time was of the number. There were three ladies also, whose great vivacity lent a zest to the lively conversation which soon followed on the departure of the stage. I took a seat with the driver who, to my great joy, happened to be the world-renowned Jabe Sawin, that prince of stage drivers and storytellers. The usual number of idlers stayed around the door of the hotel to see that everything was right, when a nod from the agent signifying that all were on board gave Jabe the signal to go ahead. With a 'chirrup' and a 'go lang,' accompanied by several artistic twirls of a long, heavy whip-lash, duly accented with those sharp exclamations so familiar to equine ears, the team started off at a rattling pace that took us speedily beyond the limits of the city and upon the main road towards Augusta. It was a bright September morning, and all the omens were favorable for a delightful day. Holmes on the inside and Jabe on the out, soon had their tongues a-wagging, but conversation did not become lively and general, until we had made the first shift of horses about 10 miles out. It was a time-honored custom of 'the fathers' on the occurrence of any 'shift,' to alight from the stage, stretch their cramped legs and take a little, well, what for shortness, we will call 'air.' We were not long in making this shift when the customary 'stage ready' was sounded and we were speeding on our way. I had exchanged seats with a gentleman who wished to indulge in a 'fragrant Havana,' and took one on the inside. The effects of the 'air' soon manifested itself in the increased volubility of the gentlemen passengers, and everything went 'merry as a marriage bell.' For genuine politeness for brilliant conversation, for whit, humor, repartee, fun, anecdote and story, commend me to the old-fashioned stagecoach forever. People are not less sociable now than formerly, I suppose, but the present methods of traveling forbid general conversation. Probably there is scarcely one of the 'old boys' living who cannot recall some pleasant incident or anecdote of the stagecoach.

In about an hour and one-half, the fun meanwhile 'growing fast and furious,' we had come to the next shifting place, where all hands again got out, 'freshened the nip,' to use a sailor's phrase, and we were off at top speed for Brunswick. I resumed my seat with the driver, who for the balance of the journey kept my verdant mouth agape with his marvelous tales. Arrived at Brunswick. We had an excellent dinner of which all ate heartily. By this time the passengers had become thoroughly acquainted with each other, and were on the best of terms. Whit, humor and laughter were in order, and if that lot of passengers never smiled again they had abundant fun that day. The village idlers as usual had gathered about the

'Tontine' (I think the hotel was called) in such numbers that one unaccustomed to travel might think an ovation was intended. The scene was a lively one. We had met the stage from Augusta for Portland, who's passengers, apparently as jolly as ourselves, joined us at the dinner table. Bows and courtesies were exchanged between the passengers as we were about to depart; the two drivers had passed the customary jokes concerning their respective sweethearts, when, with A 'go-lang' and a crack of the whip, each sped to his destination. The afternoon passed very much as the forenoon had, and to relate just what was said and done would be a useless reputation. Suffice it to say, we arrived safely at Augusta at 6:00 PM after a most enjoyable ride of eleven hours from Portland.

There were two classes of men then connected with staging which, though now almost obsolete, contributed so largely to the comfort and entertainment of the travelers, that they deserve a passing notice. I mean the village inn keeper and the stage driver. It was the former who, as a stage drove up in front of his hotel, stood, with a ready, happy, smiling face, ready to 'welcome the coming, or speed the departing guest,' and who with a deferential bow and gentle wave of the hand, such as Chesterfield himself might have applauded, led the way to his well warmed or well ventilated apartments, as the caprice of the weather might suggest, offering to each guest the hospitality of his house with such blandness of manner as not only to put one at ease with one's self, but to convey the impression that he was an especial object of favor and indulgence. This really polite class of old-fashioned country landlords, as then existed on all the great lines of travel, has well-nigh become extinct, but the remembrance of their hospitality and good cheer will be cherished as long as any of the 'old guard' live to tell the story.

The old stage driver belonged to a class *sui generis*. Their ways were peculiar, their dress peculiar and their language peculiar. They were the oracles of the bar-room, and were deferred to on all matters pertaining to the stable, or the road. I suppose there were not many boys fifty years ago who had not had some time felt that the climax of human greatness was only reached by the stage driver. The occupation was one of exposure, excitement and responsibility. The men were selected by their employers on account of their intelligence, honesty and general good behavior. They were the conductors, engineers and express men of their day. They not only had charge of the mails, but were often entrusted with large sums of money by banks and merchants. They also did all the errands along the road, such as buying gimcracks for the girls, drugs for the sick and balms for the aged. They were uniformly gallant to the opposite sex and mothers could trust their daughters with them for a journey with a perfect assurance of decorous treatment; although exposed to great social temptations, and though in the habit of occasionally taking a glass, they were not as a class intemperate, and I cannot recall an instant where one of the fraternity lost his situation by means of drunkenness or dishonesty. The old lot, whom I remember, have nearly all passed away. There was Twitchell, Bodge, Black, Washburn, Erastus Richards, Nat Hutchins and Jabe Sawin. Of these I do not know whether any survive. Poor old Jabe was the last I remember to have seen. Thirty odd years after the occurrence of the incidents herein related, I attended the funeral in California of this remarkable stage driver's wife, and met him there, I think, the last time. He is now dead. He played his role in the grand drama of life, humble though it was, with consummate skill, causing a thousand smiles to one tear, and died leaving behind him a host of friends who remember him with kindly feeling.[19]

Stage Humor

As early as 1771, stage coaches were the basis for humor and lesson teaching. Here is a good example:

A Fable to be Studied by Modern Patriots and Politicians.

The weather was sultry, and the roads full of dust, when a stagecoach came up with a traveler, who was toiling and sweating on foot, 'For heaven's sake, (said he to the driver,) give me a place; no matter where, whether within or without, on the coach box, behind the coach, take pity on me, or herewith I die with fatigue.'

'Every corner is already crammed,' (answered the driver,) 'You cannot be admitted (cried the passengers) unless you mean to have us all stifled.' 'Alas! (Said the traveler) but now I examine things more closely, I would not accept of a place. To show you, however, how disinterested I am, and that I can return good for evil, I will warn you of your danger. The coachman is drunk, the horses unruly, and the axle tree cracked.' At first, they laughed at him, and disregarded his warning, but he called so loud, and persisted so obstinately in his assertion, that one of the passengers, more timid than the rest, thought they might at least examine the matter, and so stepped out. The driver who had assured them that everything was safe and in good order, was disobliged; and the crafty traveler, watching the opportunity, jumped in the vacant seat, 'drive on,' says he to the coachman; 'and give me leave to assure you, (addressing himself to his fellow passengers) this is the soberest coachman on the road, his horses quieted, and his axle-tree heart of oak.'[20]

A Letter from Mr. Andrews.

On a late return from a journey into the eastern country, I was much amused with a lively chat of a civil, good natured Irishman, who happened to be

[19] Maine Farmer (Augusta, Maine) November 11, 1876 p4.

[20] New Hampshire Gazette (Portsmouth, New Hampshire) June 7, 1771 p3.

my fellow traveler in the mail stage. On our arrival at Portland, in the evening, the driver informed us that we must be ready, that he should call for us precisely at four the next morning. In the morning when I came down, a little before the hour, I found my good-natured companion all awake and ready to step into the stage. I seated myself with him and enjoyed his wit and merriment in the bar room until the clock told five. The landlord then came down, rubbing his eyes. 'what,' says he, 'not gone yet; hasn't the stage called for you?' 'Fath (Fathom), and indeed, sir,' says my companion, 'it has not, and I believe we'll be here still.' 'then I suspect,' says the landlord, 'that you are left behind, or that the driver has overslept himself; but I will send and see.' Just at this moment up comes the stage full speed to the door. 'Step in, step in, gentleman,' exclaims the driver, 'I am more than an hour too late; I must meet the returning stage at Saco, if I kill my horses; do be quick gentleman.' We stepped in; the door was shut; smack went the whip, round went the wheels and in a trice we arrived at the halfway house between Portland and Saco; when, one of the hind horses fell lifeless to the ground. We were out very quick, and went around to look at the fallen horse, when Paddy, putting his finger upon the distended eyeball of the poor creature, and finding it motionless, lifted his hands and exclaimed: 'Well, by — , I have no doubt this poor horse has been dead for miles back, but he could not stop to lay down.'[21]

[21] Maine Gazette (Bath, Maine) October 10, 1823 p5.

A Pleasant Stagecoach Companion.

A remarkably tall man traveling inside of a stagecoach, greatly incommoded the occupant of the opposite seat, by the disposition of its feet. For many stages the sufferer bore his state with heroic fortitude, and no word of complaint escaped his lips, until the coach again stopped to change horses, when the tall gentleman unfastening the door exclaimed, 'Well, I should just get out and stretch my legs a little.' 'Don't for God's sake don't (replied the other) they're too long already!'[22]

Anecdote.

A stage driver the other day, between Bangor and this place, was hailed by a woman, and requested to take a bed stead and its accompanying bed and clothing on board, together with her dear self. This was a poseur for John's gallantry, (for stage drivers are always ready to oblige the ladies,) but making her a very civil bow, he said, 'I should like to oblige ye, ma'am very much, but I've engaged to take a double saw mill aboard a mile or two ahead, and as you might find it an uncomfortable companion, I should advise you to wait a week or so till the mill-moving business is over.'[23]

New England Stage Drivers.

It may not be known to many of our readers (says a Southern Paper) that stage driving in New England is a profession, and a crack profession it is

[22] Eastern Argus (Portland, Maine) September 20, 1831.

[23] Ibid April 18, 1837 p1.

too. The gentlemen of the whip form a very numerous and distinct class, which ranks scarcely below any other. The New England stage driver is quite a different sort of person from what you meet with in the same pursuit in any other section of the country; gentlemanly, intelligent, of superior business habits, (so much so that he is frequently entrusted with large sums of money to make purchases for dealers) well dressed, often the proprietor of the horses he drives, always with a plenty of money for his own uses, and a universal favorite on his route. When the hour of starting arrives, his coach is driven to the door, he mounts the box, takes the ribands in his gloved hands, cracks his whip, and dashes through his route, drops his reigns in the ostler's hands, and gives himself no further concern of horses or coach until they are driven to the door again, for his return passage on the following day. In the meantime, he plays the gentlemen. We remember one of these, a clever fellow who formerly drove into Concord New Hampshire. He prided himself on the fleetness of his horses, which were his own. A Southerner happened one day among his passengers. Unaccustomed to his rapid rate of traveling, he remonstrated against what he considered the drivers cruelty to his horses. The driver said nothing, but dashed onward. The Southerner insisted. The driver replied not, but knowing the mettle of his team, cracked up the leaders. The Southerner grew furious, and finally swore terrible oaths that he would report him to the proprietors as unfit for his place. 'My dear fellow,' said the driver, at last, 'you have waked up the wrong passenger. The horses are my own;' and to show him that some things could be done as well as others, put them to a still quicker speed, and whirled the astonished passenger into Concord at a rate he had never imagined could be traveled in a stagecoach.[24]

[24] Maine Cultivator and Hallowell gazette (Hallowell, Maine) February 21, 1840 p4.

Conclusion

The period from the Revolutionary War to the arrival of the railroad was an important time in the formation of what Maine would become. Travel for commerce and government participation, as well as pleasure, became a necessity and the stagecoach evolved to fill that need. From open wagons to elaborate Concord stages, the traveling public could find public transportation available as close as the local tavern or inn. Of course, this did not happen overnight. It took years of planning and hard work, building roads, locating stage stops, and setting up the logistics to enable stage owners to provide horse power and reasonable comfort and safety for its passengers. The stagecoach era began slowly in the late 18th century, building up to its peak in the early 19th.

Once the operations were established, owners usually did not drive their own rigs over the routes. For this they hired stage drivers. These men, at least in New England, were the cream of the crop. They had to be. On their coach, they were the captain of the ship. Many even handled large sums of cash or negotiable papers, fully trusted to convey them to the final destination. They would even take responsibility for transporting minor children without their parents. Many drivers moved on to positions in the railroads when the stages ceased to be.

The stage and tavern owners were always highly respected in their community. They would participate in government or charities, always deeply involved in their civic obligations. They were also astute business men, even though they might not always get rich from their chosen professions.

Finally, the whole industry came to a stop with the arrival of the railroads. Even that did not happen all at once. Stage lines would move their point of origin to meet railroads at their "end of line." As the railroads covered more territory, the stagecoaches covered less, but until the beginning of the twentieth century, they were still in demand. Today, they show up at county fairs or in museums, but they will still cause a stir in the heart of those who can picture a whole town waiting for the arrival of a daily stage, to learn the news of the world and to see what strangers might be coming to visit.

The men and women of the stagecoach industry exemplify the principal that hard work, participation in community, and dedication to family leads to a successful and meaningful life.

Color Image Credits

Title	Artist / Photographer	Courtesy
State House Boston	James Brown Marston	Mass. Historical Society
State House Portland	Anna M. Buknam	Maine Historical Society
State House Augusta	Charles Codman	Maine State Museum
Merrimack Hotel	Unknown	Postcard-Public Domain
Earl of Halifax Tavern	Sarah Haven Foster	Portsmouth City Library
Stoodley Tavern	L. J. Hanchett	L. J. Hanchett
William Pitt Tavern Parlor	"	Strawberry Banke Museum
Rice Tavern	Sarah Haven Foster	Portsmouth City Library
Barnard Tavern	Unknown	John McInnis Auctioneers
Buxton Tavern	L. J. Hanchett	L. J. Hanchett
Peacock Tavern	"	"
Cushman Tavern	"	"
Bachelder Tavern	"	"
Hathorn Taven	"	"
Yellow Concord Stage Coach	C. Peter James	C. Peter James
Lambert Tavern Sign	Unknown	David A. Schorsch
Washington Hotel Sign	Unknown	Peabody Essex Museum
Merrill Hotel Wallpaper	Tia Wilson	Durham Historical Society
Shepard Tavern Parlor	Unknown	Boston Museum of Fine Arts
Bath Hotel	Unknown	Postcard-Public Domain
First American MacAdam Rd.	Carl Rakeman	Public Domain
American Stagecoach 1827	Capt. Basil Hall, R. N.	Public Domain

Bibliography

Allen, Charles Edwin. *History of Dresden*, Maine. 1931.

Anderson, Hayden L. V. *Canals and Inland Waterways of Maine*. Maine Historical Society, 1982.

Bourne, Edward E., and E. E. Bourne, Jr.. *The History of Wells and Kennebunk from the Earliest Settlement to the Year 1820: At Which Time Kennebunk Was Set Off, and Incorporated*. Higginson Book, 1997.

Brewster, Charles W. *Rambles about Portsmouth. Principally from Tradition and Unpublished Documents*. Portsmouth, L.W. Brewster, 1873.

Brown, Warren. *History of the Town of Hampton Falls, New Hampshire: from the Time of the First Settlement within Its Borders, 1640 until 1900*. Printed by the J.R. Clarke Co., 1900.

Burton, Lance W. *The Wigwam Resort*. Arcadia Pub., 2007.

Chase, Fannie S. *Wiscasset in Pownalborough*. The Southworth-Anthoensen Press, 1941.

Clason, Oliver Barrett. *History of Litchfield Maine: And an Account of Its Centennial Celebration, 1895*. Heritage Books, 2010.

Clayton, W. W. *History of Cumberland County, Maine*. Philadelphia, Everts and Peck, 1880.

Cochrane, Harry H. *History of Monmouth and Wales*. Higginson Book Co., 1997.

Corliss, Augustus Whittemore. *Old Times of North Yarmouth, Maine: A Facsimile of the Thirty-Two Quarterly Magazines Edited and Published by Augustus W. Corliss as Well as the Single Issue of The Wescustogo Chronicle, 1877-1885*. New Hampshire Publishing Co., 1977.

Currier, John J. *History of Newburyport, Massachusetts*, 1764-1909. Currier, 1906.

Crawford, Mary Caroline. *Among Old New England Inns: Being an Account of Little Journeys to Various Quaint Inns and Hostelries of Colonial New England*. Heritage Books, 1998.

Danforth-Byras, Karen, and Hayden L. V. Anderson, editors. *Litchfield Yesterdays: A Short Historical Tour and Glimpse of Bygone Days.* Compiled and Edited by the Litchfield Heritage Committee and the Historical Society of Litchfield, 1976.

Devine, Marge. "Lambert's Tavern: C 1741." *Falmouth History* 4.1, 2003.

Dole, Samuel Thomas. *Windham in the Past.* Merrill and Weber, 1916.

Dow, Joseph. *History of the Town of Hampton, New Hampshire: from Its Settlement in 1638, to the Autumn ... of 1892* (Classic Reprint). Vols. I & II, Forgotten Books, 2016.

Duncan, John M. *Travels Through Part of the United States and Canada in 1818 and 1819.* Forgotten Books, 2016.

Dow, George Francis. *History of Topsfield, Massachusetts.* Merriam-Eddy, 1982.

Earle, Alice Morse. *Stage-coach and Tavern Days.* Macmillan, 1905.

Elwell, Edward Henry. *Successful Business Houses of Portland.* Nabu, 2012.

Essex Institute. *The Essex Institute Historical Collections Vol. XLII 1906.* Essex Institute, 1906.

Fagan, John T. *The Charter and Ordinances of the City of Portland: Together with Acts of the Legislature, relating to the City, Collated and Revised by Authority of a Joint Committee of the City Council.* Portland, Me., Foster City, 1868.

Field, Edward. *The Colonial Tavern.* Heritage Books, 1989.

Folsom, George. *History of Saco and Biddeford.* New Hampshire Publishing, 1975.

Forbes, Allan, and Ralph M. Eastman. *Taverns and Stagecoaches of New England; Anecdotes and Tales Recalling the Days of Stagecoach Travel and the Ancient Hostelries Where Strangers Tarried.* State Street Trust, 1953.

Forbes, Allan, and Ralph M. Eastman. *Taverns and Stagecoaches of New England; Anecdotes and Tales Recalling the Days of Stagecoach Travel and the Ancient Hostelries Where Strangers Tarried.* Vol. II: State Street Trust, 1954.

Garvin, Donna-Belle, and James L. Garvin. *On the Road North of Boston: New Hampshire Taverns and Turnpikes, 1700-1900.* University of New England, 2003.

Gillespie, W. M. *A Manual of the Principles and Practice of Road-Making.* New York, A. S. Barnes & CO., 1853.

Gilpatric, George A. *The Village of Kennebunk, Maine: Interesting Facts from Old Documents and Maps, and Observations by the Author.* Star, 1935.

Goold, Nathan, and William Willis. *Alice Greele's Tavern.*

Griggs, Frank, Jr. "Newburyport Bridge." *Structure Magazine* June 2013.

Hall, Basil. *Forty Etchings from Sketches Made with a Camera Lucida in North America, in 1827 and 1828*. London, 1830.

Harnedy, Jim. *Around Wiscasset: Alna, Dresden, Westport Island, Wiscasset, and Woolwich*. Arcadia Pub., 2009.

Hazlett, C. A. *The Portsmouth Book*: Higginson Book, 2000.

Holmes, Oliver Wendell, and Peter Thomas. Rohrbach. *Stagecoach East: Stagecoach Days in the East from the Colonial Period to the Civil War*. Smithsonian Institution, 1983.

Jones, Herbert G. *The King's Highway from Portland to Kittery: Stagecoach & Tavern Days on the Old Post Road*. Longfellow, 1953.

Kingsbury, Henry D., et al. *Illustrated History of Kennebec County, Maine: 1625, 1799, 1892*. New York, H.W. Blake & Co., 1892.

Kopell, Jonathan C. S. *Origins of Shareholder Advocacy*. Palgrave Macmillan, 2016.

Kyle, George Alexander. *The Straight Road: A Short Account of the Newburyport Turnpike and Early Days in Everett, Massachusetts*. Everett National Bank, 1927.

Laskowski, Amy. "Secrets of the Three Cranes Tavern." *Boston University DIG*. 9 Nov. 2014.

Lassiter, Cheryl A. *A Meet and Suitable Person. Tavernkeeping in Old Hampton, New Hampshire, 1638-1783*. Blue Petal Press, 2012.

Latham, Ettie J. *History of the Town of Pownal: Written for Its Centennial, Sept. 2, 1908*. Heirloomsreunited.com, 2012.

Litchfield, P.W. *Industrial Voyage*. Doubleday, 1954.

Little, George Thomas. *General Catalog of Bowdoin College and Medical School 1794*.

Longfellow, Henry Wadsworth. *The Poetical Works of Henry Wadsworth Longfellow*. Houghton Mifflin and Co., 1884.

The Maine Register and United States' Almanac for the Year of Our Lord 1820 ... : Containing Civil, Judicial, Ecclesiastical and Military Lists in Maine, Associations and Corporate Institutions ... Names of the Post Towns and Post Masters ... : Also, the Constitution as Reported by the Convention, Catalogues of the Officers of the General Government ... times of the Sittings of the Several Courts, and a Variety of Other Interesting Articles. Portland, F. Douglas & A. Shirley, 1819.

Maher, Frances Soule and Burnette Bailey Wallace. *History of Woolwich, Maine a Town Remembered*. Woolwich Historical Society, 1994.

Merrill, Georgia Drew, editor. *History of Androscoggin County, Maine*. Boston, W. A. Ferguson & CO., 1891.

Marlowe, George F. *Coaching Roads of Old New England: Their Inns and Taverns and Their Stories*. Macmillan, 1945.

Moore, Christopher. *The Loyalists Revolution, Exile, Settlement*. McClelland & Stewart, 1994.

Mower, Walter Lindley. *Sesquicentennial History of the Town of Greene, Androscoggin County, Maine, 1775-1900, with Some Matter Extending to a Later Date*. 1938.

Nason, Emma Huntington. *Old Hallowell on the Kennebec*. Sam Teddy Pub., 2009.

North, James W. *The History of Augusta*. Sam Teddy, 2016.

Nowers, Elizabeth P. *Portsmouth Women: Madams & Matriarchs Who Shaped New Hampshire's Port City*. Arcadia Publishing, 2013.

Nutting, Wallace. *Massachusetts Beautiful*. Bonanza Books, 1980.

Official Register of the United States: The Post-Office Department and the Postal Service. G.P.O., 1839.

Owen, Henry Wilson, and Edward Clarence Plummer. *The Edward Clarence Plummer History of Bath, Maine*. Bath Area Bicentennial Committee, 1976.

Parks, Roger N. *Roads and Travel in New England: 1790 – 1840*. Old Sturbridge Village, 1967.

Reed, Parker McCobb. *History of Bath and Environs: Sagadahoc County, Maine, 1607-1894*. Forgotten Books, 2015.

Ring, Elizabeth. *Aids to the Teaching of Maine in the Public Schools*. Fred L. Tower.

Ripley, Hubert George. *New England Inns and Taverns*. R.F. Whitehead, 1931.

Rowe, William Hutchinson. *Ancient North Yarmouth and Yarmouth, Maine, 1636-1936: A History*: Picton Press, 2000.

Stackpole, Everett S. *History of Durham, Maine: With Genealogical Notes*. Forgotten Books, 2016.

Stackpole, Everett Schermerhorn, and David Colby Young. *History of Winthrop, Maine: with Genealogical Notes*. Heritage Books, 1994.

State Street Trust Co. *Some Events of Boston and Its Neighbors*: Walton Advertising, 1917.

Sylvester, Herbert Milton. *Olde Cascoe: Ye Romance of Casco Bay*. W.B. Clarke, 1909.

Thurston, David. *A Brief History of Winthrop: from 1764 to October 1855*. Sam Teddy Publishing, 2011.

Thurston, Florence G., and Harmon S. Cross. *Three Centuries of Freeport, Maine*. Higginson Book, 1998.

The Historical Collections of the Topsfield Historical Society. The Society, 1902.

Twining, Thomas. *Travels in America 100 Years Ago*. Harper, 1902.

Wallace, Charlotte, and Donald Wallace. *A Story of Falmouth, Maine*. 1976.

Wheeler, George Augustus., and Henry Warren. Wheeler. *History of Brunswick, Topsham, and Harpswell, Maine: Including the Ancient Territory Known as Pejepscot*. Boston, Alfred Mudge and Sons, 1878

Weld, Isaac. *Travels through the States of North America, and the Provinces of Upper and Lower Canada, during the Years 1795, 1796, and 1797.* London: ed for John Stockdale ... , 1807.

Wellman, Thomas B. *History of the Town of Lynnfield, Mass., 1635-1895.* Heritage Books, 1997.

Willis, J. L. M. *Old Eliot; a Quarterly Magazine of the History and Biography of the Upper Parish of Kittery, Now Eliot.* A. Caldwell, 1897.

Willis, J. L. M. *The Submission of Maine to Massachusetts: Signed at William Everett's Tavern, in Kittery, Now Eliot Maine, November 16, 1652: With Brief Sketches of the Signers.* Historical, 1909.

Willis, William. *The History of Portland.* New Hampshire Publishing, 1972.

Willis, William. *Business Directory of the Subscribers to the New Map of Maine: with a Brief History and Description of the State.* Forgotten Books, 2016.

Wood, Frederic J. *The Turnpikes of New England*: Branch Line, 1997.

Index

C

D

E

F

G

H

I

J

K

L

M

I

J

K

L

M

N

O

P

R

S

T

V

W